Meta-Ana

An Updated Collection from the Stata J

Meta-Analysis in Stata:

An Updated Collection from the Stata Journal

JONATHAN A. C. STERNE, collection editor
Department of Social Medicine
University of Bristol
Bristol, UK

H. JOSEPH NEWTON, *Stata Journal* editor
Department of Statistics
Texas A&M University
College Station, TX

NICHOLAS J. COX, *Stata Journal* editor
Department of Geography
Durham University
Durham City, UK

A Stata Press Publication
StataCorp LP
College Station, Texas

Copyright © 2009 by StataCorp LP
All rights reserved. First edition 2009

Published by Stata Press, 4905 Lakeway Drive, College Station, Texas 77845
Typeset in LaTeX 2_ε
Printed in the United States of America

10 9 8 7 6 5 4 3 2 1

ISBN-10: 1-59718-049-1
ISBN-13: 978-1-59718-049-8

Contents

Introduction   vii

Install the software   xi

1 Meta-analysis in Stata: metan, metacum, and metap   1

metan—a command for meta-analysis in Stata ....
.... M. J. Bradburn, J. J. Deeks, and D. G. Altman   3

metan: fixed- and random-effects meta-analysis ....
... R. J. Harris, M. J. Bradburn, J. J. Deeks, R. M. Harbord, D. G. Altman,
and J. A. C. Sterne   29

Cumulative meta-analysis .... J. A. C. Sterne   55

Meta-analysis of p-values .... A. Tobias   65

2 Meta-regression: metareg   69

Meta-regression in Stata .... R. M. Harbord and J. P. T. Higgins   70

Meta-analysis regression .... S. Sharp   97

3 Investigating bias in meta-analysis: metafunnel, confunnel, metabias, and metatrim   107

Funnel plots in meta-analysis .... J. A. C. Sterne and R. M. Harbord   109

Contour-enhanced funnel plots for meta-analysis ....
.... T. M. Palmer, J. L. Peters, A. J. Sutton, and S. G. Moreno   124

Updated tests for small-study effects in meta-analyses ....
.... R. M. Harbord, R. J. Harris, and J. A. C. Sterne   138

Tests for publication bias in meta-analysis .... T. J. Steichen   151

Tests for publication bias in meta-analysis ....
.... T. J. Steichen, M. Egger, and J. A. C. Sterne   162

Nonparametric trim and fill analysis of publication bias in meta-analysis ....
.... T. J. Steichen   165

4 Advanced methods: metandi, glst, metamiss, and mvmeta   179

metandi: Meta-analysis of diagnostic accuracy using hierarchical logistic regression..

......................................R. M. Harbord and P. Whiting   181

Generalized least squares for trend estimation of summarized dose–response data

...........................N. Orsini, R. Bellocco, and S. Greenland   200

Meta-analysis with missing data..............I. R. White and J. P. T. Higgins   218

Multivariate random-effects meta-analysis.........................I. R. White   231

Appendix   249

Author index   253

Command index   259

Introduction

This first collection of articles from the *Stata Technical Bulletin* and the *Stata Journal* brings together updated user-written commands for meta-analysis, which has been defined as a statistical analysis that combines or integrates the results of several independent studies considered by the analyst to be combinable (Huque 1988). The statistician Karl Pearson is commonly credited with performing the first meta-analysis more than a century ago (Pearson 1904)—the term "meta-analysis" was first used by Glass (1976). The rapid increase over the last three decades in the number of meta-analyses reported in the social and medical literature has been accompanied by extensive research on the underlying statistical methods. It is therefore surprising that the major statistical software packages have been slow to provide meta-analytic routines (Sterne, Egger, and Sutton 2001).

During the mid-1990s, Stata users recognized that the ease with which new commands could be written and distributed, and the availability of improved graphics programming facilities, provided an opportunity to make meta-analysis software widely available. The first command, `meta`, was published in 1997 (Sharp and Sterne 1997), while the `metan` command—now the main Stata meta-analysis command—was published shortly afterward (Bradburn, Deeks, and Altman 1998). A major motivation for writing `metan` was to provide independent validation of the routines programmed into the specialist software written for the Cochrane Collaboration, an international organization dedicated to improving health care decision-making globally, through systematic reviews of the effects of health care interventions, published in The Cochrane Library (see www.cochrane.org). The groups responsible for the `meta` and `metan` commands combined to produce a major update to `metan` that was published in 2008 (Harris et al. 2008). This update uses the most recent Stata graphics routines to provide flexible displays combining text and figures. Further articles describe commands for cumulative meta-analysis (Sterne 1998) and for meta-analysis of *p*-values (Tobias 1999), which can be traced back to Fisher (1932). Between-study heterogeneity in results, which can cause major difficulties in interpretation, can be investigated using meta-regression (Berkey et al. 1995). The `metareg` command (Sharp 1998) remains one of the few implementations of meta-regression and has been updated to take account of improvements in Stata estimation facilities and recent methodological developments (Harbord and Higgins 2008).

Enthusiasm for meta-analysis has been tempered by a realization that flaws in the conduct of studies (Schulz et al. 1995), and the tendency for the publication process to favor studies with statistically significant results (Begg and Berlin 1988; Dickersin, Min, and Meinert 1992), can lead to the results of meta-analyses mirroring overoptimistic results from the original studies (Egger et al. 1997). A set of Stata commands—`metafunnel`, `confunnel`, `metabias`, and `metatrim`—address these issues both graphically (via routines to draw standard funnel plots and "contour-enhanced" funnel plots) and statistically, by providing tests for funnel plot asymmetry, which can be used to diagnose publication bias and other small-study effects (Sterne, Gavaghan, and Egger 2000; Sterne, Egger, and Moher 2008).

This collection also contains advanced routines that exploit Stata's range of estimation procedures. Meta-analysis of studies that estimate the accuracy of diagnostic tests, implemented in the `metandi` command, is inherently bivariate, because of the trade-off between sensitivity and specificity (Rutter and Gatsonis 2001; Reitsma et al. 2005). Meta-analyses of observational studies will often need to combine dose–response relationships, but reports of such studies often report comparisons between three or more categories. The method of Greenland and Longnecker (1992), implemented in the `glst` command, converts categorical to dose–response comparisons and can thus be used to derive the data needed for dose–response meta-analyses. White and colleagues (White and Higgins 2009; White 2009) have recently provided general routines to deal with missing data in meta-analysis, and for multivariate random-effects meta-analysis.

Finally, the appendix lists user-written meta-analysis commands that have not, so far, been accepted for publication in the *Stata Journal*. For the most up-to-date information on meta-analysis commands in Stata, readers are encouraged to check the Stata frequently asked question on meta-analysis:

http://www.stata.com/support/faqs/stat/meta.html

Those involved in developing Stata meta-analysis commands have been delighted by their widespread worldwide use. However, a by-product of the large number of commands and updates to these commands now available has been that users find it increasingly difficult to identify the most recent version of commands, the commands most relevant to a particular purpose, and the related documentation. This collection aims to provide a comprehensive description of the facilities for meta-analysis now available in Stata and has also stimulated the production and documentation of a number of updates to existing commands, some of which were long overdue. I hope that this collection will be useful to the large number of Stata users already conducting meta-analyses, as well as facilitate interest in and use of the commands by new users.

Jonathan A. C. Sterne
February 2009

1   References

Begg, C. B., and J. A. Berlin. 1988. Publication bias: A problem in interpreting medical data. *Journal of the Royal Statistical Society, Series A* 151: 419–463.

Berkey, C. S., D. C. Hoaglin, F. Mosteller, and G. A. Colditz. 1995. A random-effects regression model for meta-analysis. *Statistics in Medicine* 14: 395–411.

Bradburn, M. J., J. J. Deeks, and D. G. Altman. 1998. sbe24: metan—an alternative meta-analysis command. *Stata Technical Bulletin* 44: 4–15. Reprinted in *Stata Technical Bulletin Reprints*, vol. 8, pp. 86–100. College Station, TX: Stata Press. (Updated article is reprinted in this collection on pp. 3–28.)

Dickersin, K., Y. I. Min, and C. L. Meinert. 1992. Factors influencing publication of research results: Follow-up of applications submitted to two institutional review boards. *Journal of the American Medical Association* 267: 374–378.

Egger, M., G. Davey Smith, M. Schneider, and C. Minder. 1997. Bias in meta-analysis detected by a simple, graphical test. *British Medical Journal* 315: 629–634.

Fisher, R. A. 1932. *Statistical Methods for Research Workers*. 4th ed. London: Oliver & Boyd.

Glass, G. V. 1976. Primary, secondary, and meta-analysis of research. *Educational Researcher* 10: 3–8.

Greenland, S., and M. P. Longnecker. 1992. Methods for trend estimation from summarized dose–reponse data, with applications to meta-analysis. *American Journal of Epidemiology* 135: 1301–1309.

Harbord, R. M., and J. P. T. Higgins. 2008. Meta-regression in Stata. *Stata Journal* 8: 493–519. (Reprinted in this collection on pp. 70–96.)

Harris, R. J., M. J. Bradburn, J. J. Deeks, R. M. Harbord, D. G. Altman, and J. A. C. Sterne. 2008. metan: fixed- and random-effects meta-analysis. *Stata Journal* 8: 3–28. (Reprinted in this collection on pp. 29–54.)

Huque, M. F. 1988. Experiences with meta-analysis in NDA submissions. *Proceedings of the Biopharmaceutical Section of the American Statistical Association* 2: 28–33.

Pearson, K. 1904. Report on certain enteric fever inoculation statistics. *British Medical Journal* 2: 1243–1246.

Reitsma, J. B., A. S. Glas, A. W. S. Rutjes, R. J. P. M. Scholten, P. M. Bossuyt, and A. H. Zwinderman. 2005. Bivariate analysis of sensitivity and specificity produces informative summary measures in diagnostic reviews. *Journal of Clinical Epidemiology* 58: 982–990.

Rutter, C. M., and C. A. Gatsonis. 2001. A hierarchical regression approach to meta-analysis of diagnostic test accuracy evaluations. *Statistics in Medicine* 20: 2865–2884.

Schulz, K. F., I. Chalmers, R. J. Hayes, and D. G. Altman. 1995. Empirical evidence of bias. Dimensions of methodological quality associated with estimates of treatment effects in controlled trials. *Journal of the American Medical Association* 273: 408–412.

Sharp, S. 1998. sbe23: Meta-analysis regression. *Stata Technical Bulletin* 42: 16–22. Reprinted in *Stata Technical Bulletin Reprints*, vol. 7, pp. 148–155. College Station, TX: Stata Press. (Reprinted in this collection on pp. 97–106.)

Sharp, S., and J. A. C. Sterne. 1997. sbe16: Meta-analysis. *Stata Technical Bulletin* 38: 9–14. Reprinted in *Stata Technical Bulletin Reprints*, vol. 7, pp. 100–106. College Station, TX: Stata Press.[1]

Sterne, J. 1998. sbe22: Cumulative meta analysis. *Stata Technical Bulletin* 42: 13–16. Reprinted in *Stata Technical Bulletin Reprints*, vol. 7, pp. 143–147. College Station, TX: Stata Press. (Updated article is reprinted in this collection on pp. 55–64.)

Sterne, J. A. C., M. Egger, and D. Moher. 2008. Addressing reporting biases. In *Cochrane Handbook for Systematic Reviews of Interventions*, ed. J. P. T. Higgins and S. Green, 297–334. Chichester, UK: Wiley.

Sterne, J. A. C., M. Egger, and A. J. Sutton. 2001. Meta-analysis software. In *Systematic Reviews in Health Care: Meta-Analysis in Context*, 2nd edition, ed. M. Egger, G. Davey Smith, and D. G. Altman, 336–346. London: BMJ Books.

Sterne, J. A. C., D. Gavaghan, and M. Egger. 2000. Publication and related bias in meta-analysis: Power of statistical tests and prevalence in the literature. *Journal of Clinical Epidemiology* 53: 1119–1129.

Tobias, A. 1999. sbe28: Meta-analysis of p-values. *Stata Technical Bulletin* 49: 15–17. Reprinted in *Stata Technical Bulletin Reprints*, vol. 9, pp. 138–140. College Station, TX: Stata Press. (Updated article is reprinted in this collection on pp. 65–68.)

White, I. R. 2009. Multivariate random-effects meta-analysis. *Stata Journal*. Forthcoming. (Preprinted in this collection on pp. 231–247.)

White, I. R., and J. P. T. Higgins. 2009. Meta-analysis with missing data. *Stata Journal*. Forthcoming. (Preprinted in this collection on pp. 218–230.)

---

1. The original command to perform meta-analysis was `meta`, documented in the sbe16 articles; `meta` is now `metan`. `metan` is described in an updated article, sbe24, on pages 3–28 of this collection.—Ed.

Install the software

You can obtain all the user-written commands that are described in this collection from within Stata. Download the installation command by typing

```
. net from http://www.stata-press.com/data/mais
. net install mais
```

After installing this file, type `spinst_mais` to obtain all the user-written commands that are discussed in this collection, except for those commands listed in the appendix. Instructions on how to obtain those commands are given in the appendix. If there are any error messages after typing `spinst_mais`, follow the instructions at the bottom of the output to complete the download.

Part 1

Meta-analysis in Stata: metan, metacum, and metap

The `metan` command is the main Stata meta-analysis command. In its latest version, it provides highly flexible facilities for doing meta-analyses and graphing their results. Its worldwide use testifies to the dedication and skills of Michael Bradburn, who did most of the original programming and then added a range of facilities in response to user requests, and Ross Harris, who redesigned the graphics and updated them to Stata 9 and added further options.

`metan` is described in two articles. The first—by Bradburn, Deeks, and Altman—was published in the *Stata Technical Bulletin* in 1998. For this collection, it has been updated to describe and use the most recent `metan` syntax, with the graphics also having been updated. Editorial notes explain where other commands originally published and distributed with `metan` have now been superseded. The additional facilities made available since the publication of the original `metan` command are described in the 2008 *Stata Journal* article by Harris et al.

The evolution of evidence over time can be described and displayed using cumulative meta-analysis. `metacum`—a command for cumulative analysis—was described by Sterne in the *Stata Technical Bulletin* in 1998. For this collection, the `metacum` command was updated by Ross Harris to use version 9 graphics and the same syntax as `metan`, and the original article has been updated to reflect this.

In some circumstances, only the p-value from each study is available. Meta-analysis of p-values, implemented in the `metap` command (Tobias 1999), can be traced back to Fisher (1932). However, users should be aware that such analyses ignore the direction of the effect in individual studies and so are best seen as providing an overall test of the null hypothesis of no effect.

1 References

Bradburn, M. J., J. J. Deeks, and D. G. Altman. 1998. sbe24: metan—an alternative meta-analysis command. *Stata Technical Bulletin* 44: 4–15. Reprinted in *Stata Tech-*

nical Bulletin Reprints, vol. 8, pp. 86–100. College Station, TX: Stata Press. (Updated article is reprinted in this collection on pp. 3–28.)

Fisher, R. A. 1932. *Statistical Methods for Research Workers*. 4th ed. London: Oliver & Boyd.

Harris, R. J., M. J. Bradburn, J. J. Deeks, R. M. Harbord, D. G. Altman, and J. A. C. Sterne. 2008. metan: fixed- and random-effects meta-analysis. *Stata Journal* 8: 3–28. (Reprinted in this collection on pp. 29–54.)

Sterne, J. 1998. sbe22: Cumulative meta analysis. *Stata Technical Bulletin* 42: 13–16. Reprinted in *Stata Technical Bulletin Reprints*, vol. 7, pp. 143–147. College Station, TX: Stata Press. (Updated article is reprinted in this collection on pp. 55–64.)

Tobias, A. 1999. sbe28: Meta-analysis of p-values. *Stata Technical Bulletin* 49: 15–17. Reprinted in *Stata Technical Bulletin Reprints*, vol. 9, pp. 138–140. College Station, TX: Stata Press. (Updated article is reprinted in this collection on pp. 65–68.)

The Stata Technical Bulletin (1998)
STB-44, pp. 4–15

metan—a command for meta-analysis in Stata[1]

Michael J. Bradburn
Clinical Trials Research Unit
ScHARR
Sheffield, UK
m.bradburn@sheffield.ac.uk

Jonathan J. Deeks
Unit of Public Health, Epidemiology, and Biostatistics
University of Birmingham
Birmingham, UK
j.deeks@bham.ac.uk

Douglas G. Altman
Centre for Statistics in Medicine
University of Oxford
Oxford, UK

1 Background

When several studies are of a similar design, it often makes sense to try to combine the information from them all to gain precision and to investigate consistencies and discrepancies between their results. In recent years, there has been a considerable growth of this type of analysis in several fields, and in medical research in particular. In medicine, such studies usually relate to controlled trials of therapy, but the same principles apply in any scientific area; for example in epidemiology, psychology, and educational research. The essence of meta-analysis is to obtain a single estimate of the effect of interest (effect size) from some statistic observed in each of several similar studies. All methods of meta-analysis estimate the overall effect by computing a weighted average of the studies' individual estimates of effect.

metan provides methods for the meta-analysis of studies with two groups. With binary data, the effect measure can be the difference between proportions (sometimes called the risk difference or absolute risk reduction), the ratio of two proportions (risk ratio or relative risk), or the odds ratio. With continuous data, both observed differences in means or standardized differences in means (effect sizes) can be used. For both binary and continuous data, either fixed-effects or random-effects models can be fitted (Fleiss 1993). There are also other approaches, including empirical and fully Bayesian methods. Meta-analysis can be extended to other types of data and study designs, but these are not considered here.

---

1. The original title was *metan—an alternative meta-analysis command*. The updated syntax is by Ross Harris, Centre for Infections, Health Protection Agency, London.—Ed.

As well as the primary pooling analysis, there are secondary analyses that are often performed. One common additional analysis is to test whether there is excess heterogeneity in effects across the studies. There are also several graphs that can be used to supplement the main analysis.

2   Data structure

Consider a meta-analysis of k studies. When the studies have a binary outcome, the results of each study can be presented in a 2×2 table (table 1) giving the numbers of subjects who do or do not experience the event in each of the two groups (here called intervention and control).

Table 1. Binary data

Study i; $1 \le i \le k$	Event	No event
Intervention	a_i	b_i
Control	c_i	d_i

If the outcome is a continuous measure, the number of subjects in each of the two groups, their mean response, and the standard deviation of their responses are required to perform meta-analysis (table 2).

Table 2. Continuous data

Study i; $(1 \le i \le k)$	Group size	Mean response	Standard deviation
Intervention	n_{1i}	m_{1i}	sd_{1i}
Control	n_{2i}	m_{2i}	sd_{2i}

3   Analysis of binary data using fixed-effects models

There are two alternative fixed-effects analyses. The inverse variance method (sometimes referred to as Woolf's method) computes an average effect by weighting each study's log odds-ratio, log relative-risk, or risk difference according to the inverse of their sampling variance, such that studies with higher precision (lower variance) are given higher weights. This method uses large sample asymptotic sampling variances, so it may perform poorly for studies with very low or very high event rates or small sample sizes. In other situations, the inverse variance method gives a minimum variance unbiased estimate.

The Mantel–Haenszel method uses an alternative weighting scheme originally derived for analyzing stratified case–control studies. The method was first described for the odds ratio by Mantel and Haenszel (1959) and extended to the relative risk and risk difference by Greenland and Robins (1985). The estimate of the variance of the overall odds ratio was described by Robins, Greenland, and Breslow (1986). These methods are preferable to the inverse variance method as they have been shown to be robust when data are sparse, and give similar estimates to the inverse variance method in other situations. They are the default in the `metan` command. Alternative formulations of the Mantel–Haenszel methods more suited to analyzing stratified case–control studies are available in the `epitab` commands.

Peto proposed an assumption free method for estimating an overall odds ratio from the results of several large clinical trials (Yusuf et al. 1985). The method sums across all studies the difference between the observed $\{O(a_i)\}$ and expected $\{E(a_i)\}$ numbers of events in the intervention group (the expected number of events being estimated under the null hypothesis of no treatment effect). The expected value of the sum of $O - E$ under the null hypothesis is zero. The overall log odds-ratio is estimated from the ratio of the sum of the $O - E$ and the sum of the hypergeometric variances from individual trials. This method gives valid estimates when combining large balanced trials with small treatment effects, but has been shown to give biased estimates in other situations (Greenland and Salvan 1990).

If a study's 2×2 table contains one or more zero cells, then computational difficulties may be encountered in both the inverse variance and the Mantel–Haenszel methods. These can be overcome by adding a standard correction of 0.5 to all cells in the 2×2 table, and this is the approach adopted here. However, when there are no events in one whole column of the 2×2 table (i.e., all subjects have the same outcome regardless of group), the odds ratio and the relative risk cannot be estimated, and the study is given zero weight in the meta-analysis. Such trials are included in the risk difference methods as they are informative that the difference in risk is small.

4　Analysis of continuous data using fixed-effects models

The weighted mean difference meta-analysis combines the differences between the means of intervention and control groups ($m_{1i} - m_{2i}$) to estimate the overall mean difference (Sinclair and Bracken 1992, chap. 2). A prerequisite of this method is that the response is measured in the same units using comparable devices in all studies. Studies are weighted using the inverse of the variance of the differences in means. Normality within trial arms is assumed, and between trial variations in standard deviations are attributed to differences in precision, and are assumed equal in both study arms.

An alternative approach is to pool standardized differences in means, calculated as the ratio of the observed difference in means to an estimate of the standard deviation of the response. This approach is especially appropriate when studies measure the same concept (e.g., pain or depression) but use a variety of continuous scales. By standardization, the study results are transformed to a common scale (standard deviation units) that facilitates pooling. There are various methods for computing the standardized study results: Glass's method (Glass, McGaw, and Smith 1981) divides the differences in means by the control group standard deviation, whereas Cohen's and Hedges' methods use the same basic approach, but divide by an estimate of the standard deviation obtained from pooling the standard deviations from both experimental and control groups (Rosenthal 1994). Hedges' method incorporates a small sample bias correction factor (Hedges and Olkin 1985, chap. 5). An inverse variance weighting method is used in all the formulations. Normality within trial arms is assumed, and all differences in standard deviations between trials are attributed to variations in the scale of measurement.

5   Test for heterogeneity

For all the above methods, the consistency or homogeneity of the study results can be assessed by considering an appropriately weighted sum of the differences between the k individual study results and the overall estimate. The test statistic has a χ^2 distribution with $k - 1$ degrees of freedom (DerSimonian and Laird 1986).

6   Analysis of binary or continuous data using random-effects models

An approach developed by DerSimonian and Laird (1986) can be used to perform random-effects meta-analysis for all the effect measures discussed above (except the Peto method). Such models assume that the treatment effects observed in the trials are a random sample from a distribution of treatment effects with a variance τ^2. This is in contrast to the fixed-effects models which assume that the observed treatment effects are all estimates of a single treatment effect. The DerSimonian and Laird methods incorporate an estimate of the between-study variation τ^2 into both the study weights (which are the inverse of the sum of the individual sampling variance and the between studies variance τ^2) and the standard error of the estimate of the common effect. Where there are computational problems for binary data due to zero cells the same approach is used as for fixed-effects models.

Where there is excess variability (heterogeneity) between study results, random-effects models typically produce more conservative estimates of the significance of the treatment effect (i.e., a wider confidence interval) than fixed-effects models. As they give proportionately higher weights to smaller studies and lower weights to larger studies than fixed-effects analyses, there may also be differences between fixed and random models in the estimate of the treatment effect.

7   Tests of overall effect

For all analyses, the significance of the overall effect is calculated by computing a z score as the ratio of the overall effect to its standard error and comparing it with the standard normal distribution. Alternatively, for the Mantel–Haenszel odds-ratio and Peto odds-ratio method, χ^2 tests of overall effect are available (Breslow and Day 1993).

8   Graphical analyses

Three plots are available in these programs. The most common graphical display to accompany a meta-analysis shows horizontal lines for each study, depicting estimates and confidence intervals, commonly called a forest plot. The size of the plotting symbol for the point estimate in each study is proportional to the weight that each trial contributes in the meta-analysis. The overall estimate and confidence interval are marked by a diamond. For binary data, a L'Abbé plot (L'Abbé, Detsky, and O'Rourke 1987) plots the event rates in control and experimental groups by study. For all data types a funnel plot shows the relation between the effect size and precision of the estimate. It can be used to examine whether there is asymmetry suggesting possible publication bias (Egger et al. 1997), which usually occurs where studies with negative results are less likely to be published than studies with positive results.

Each trial i should be allocated one row in the dataset. There are three commands for invoking the routines; `metan`, `funnel`, and `labbe`, which are detailed below.

9   Syntax for metan

`metan` *varlist* [*if*] [*in*] [,

[*binary_data_options* | *continuous_data_options* | *precalculated_effect_estimates_options*]

measure_and_model_options *output_options* *forest_plot_options*]

binary_data_options

> `or rr rd fixed random fixedi randomi peto cornfield chi2 breslow nointeger cc(#)`

continuous_data_options

> `cohen hedges glass nostandard fixed random nointeger`

precalculated_effect_estimates_options

> `fixed random`

measure_and_model_options

> `wgt(`*wgtvar*`) second(`*model* | *estimates_and_description*`)`
> `first(`*estimates_and_description*`)`

output_options

> by(*byvar*) nosubgroup sgweight log eform efficacy ilevel(*#*)
> olevel(*#*) sortby(*varlist*)
> label([namevar = *namevar*], [yearvar = *yearvar*]) nokeep notable
> nograph nosecsub

forest_plot_options

> xlabel(*#*, ...) xtick(*#*, ...) boxsca(*#*) textsize(*#*) nobox nooverall
> nowt nostats counts group1(*string*) group2(*string*) effect(*string*) force
> lcols(*varlist*) rcols(*varlist*) astext(*#*) double nohet summaryonly rfdist
> rflevel(*#*) null(*#*) nulloff favours(*string* # *string*) firststats(*string*)
> secondstats(*string*) boxopt(*marker_options*) diamopt(*line_options*)
> pointopt(*marker_options* | *marker_label_options*) ciopt(*line_options*)
> olineopt(*line_options*) classic nowarning *graph_options*

10   Options for metan

10.1   binary_data_options

or pools ORs.

rr pools RRs; this is the default.

rd pools risk differences.

fixed specifies a fixed-effects model using the Mantel–Haenszel method; this is the default.

random specifies a random-effects model using the DerSimonian and Laird method, with the estimate of heterogeneity being taken.

fixedi specifies a fixed-effects model using the inverse-variance method.

randomi specifies a random-effects model using the DerSimonian and Laird method, with the estimate of heterogeneity being taken from the inverse-variance fixed-effects model.

peto specifies that the Peto method is used to pool ORs.

cornfield computes confidence intervals for ORs using Cornfield's method, rather than the (default) Woolf method.

chi2 displays a chi-squared statistic (instead of z) for the test of significance of the pooled effect size. This option is available only for ORs pooled using the Peto or Mantel–Haenszel methods.

breslow produces a Breslow–Day test for homogeneity of ORs.

nointeger allows the cell counts to be nonintegers. This option may be useful when a variable continuity correction is sought for studies containing zero cells but also may be used in other circumstances, such as where a cluster-randomized trial is to be incorporated and the "effective sample size" is less than the total number of observations.

cc(#) defines a fixed-continuity correction to add where a study contains a zero cell. By default, metan8 adds 0.5 to each cell of a trial where a zero is encountered when using inverse-variance, DerSimonian and Laird, or Mantel–Haenszel weighting to enable finite variance estimators to be derived. However, the cc() option allows the use of other constants (including none). See also the nointeger option.

10.2   continuous_data_options

cohen pools standardized mean differences by the Cohen method; this is the default.

hedges pools standardized mean differences by the Hedges method.

glass pools standardized mean differences by the Glass method.

nostandard pools unstandardized mean differences.

fixed specifies a fixed-effects model using the Mantel–Haenszel method; this is the default.

random specifies a random-effects model using the DerSimonian and Laird method, with the estimate of heterogeneity being taken.

nointeger denotes that the number of observations in each arm does not need to be an integer. By default, the first and fourth variables specified (containing N_intervention and N_control, respectively) may occasionally be noninteger (see nointeger in section 10.1).

10.3   precalculated_effect_estimates_options

fixed specifies a fixed-effects model using the Mantel–Haenszel method; this is the default.

random specifies a random-effects model using the DerSimonian and Laird method, with the estimate of heterogeneity being taken.

10.4   measure_and_model_options

wgt(*wgtvar*) specifies alternative weighting for any data type. The effect size is to be computed by assigning a weight of *wgtvar* to the studies. When RRs or ORs are declared, their logarithms are weighted. This option should be used only if you are satisfied that the weights are meaningful.

second(*model* | *estimates_and_description*) specifies that a second analysis may be performed using another method: fixed, random, or peto. Users may also define their own estimate and 95% CI based on calculations performed externally to metan, along with a description of their method, in the format *es lci uci description*. The results of this analysis are then displayed in the table and forest plot. If by() is used, subestimates from the second method are not displayed with user-defined estimates for obvious reasons.

first(*estimates_and_description*) completely changes the way metan operates, as results are no longer based on any standard methods. Users define their own estimate, 95% CI, and description, as in the above option, and must supply their own weightings using wgt(*wgtvar*) to control the display of box sizes. Data must be supplied in the 2 or 3 variable syntax (*theta se_theta* or *es lci uci*) and by() may not be used for obvious reasons.

10.5   output_options

by(*byvar*) specifies that the meta-analysis is to be stratified according to the variable declared.

nosubgroup specifies that no within-group results be presented. By default, metan pools trials both within and across all studies.

sgweight specifies that the display is to present the percentage weights within each subgroup separately. By default, metan presents weights as a percentage of the overall total.

log reports the results on the log scale (valid only for ORs and RRs analyses from raw data counts).

eform exponentiates all effect sizes and confidence intervals (valid only when the input variables are log ORs or log hazard-ratios with standard error or confidence intervals).

efficacy specifies results as the vaccine efficacy (the proportion of cases that would have been prevented in the placebo group had they received the vaccination). Available only with ORs or RRs.

ilevel(*#*) specifies the coverage (e.g., 90%, 95%, 99%) for the individual trial confidence intervals; the default is $S_level. ilevel() and olevel() need not be the same. See [U] **20.7 Specifying the width of confidence intervals**.

olevel(*#*) specifies the coverage (e.g., 90%, 95%, 99%) for the overall (pooled) trial confidence intervals; the default is $S_level. ilevel() and olevel() need not be the same. See [U] **20.7 Specifying the width of confidence intervals**.

sortby(*varlist*) sorts by variable(s) in *varlist*.

label([namevar=*namevar*], [yearvar=*yearvar*]) labels the data by its name, year, or both. Either or both variable lists may be left blank. For the table display, the

overall length of the label is restricted to 20 characters. If the `lcols()` option is
also specified, it will override the `label()` option.

`nokeep` prevents the retention of study parameters in permanent variables (see *Variables generated* below).

`notable` prevents the display of a results table.

`nograph` prevents the display of a graph.

`nosecsub` prevents the display of subestimates using the second method if `second()` is
used. This option is invoked automatically with user-defined estimates.

10.6   forest_plot_options

`xlabel(#,...)` defines x-axis labels. This option has been modified so that any number
of points may be defined. Also, checks are no longer made as to whether these points
are sensible, so the user may define anything if the `force` option is used. Points must
be comma separated.

`xtick(#,...)` adds tick marks to the x axis. Points must be comma separated.

`boxsca(#)` controls box scaling. This option has been modified so that the default is
`boxsca(100)` (as in 100%) and the percentage may be increased or decreased (e.g.,
80 or 120 for 20% smaller or larger, respectively).

`textsize(#)` specifies the font size for the text display on the graph. This option has
been modified so that the default is `textsize(100)` (as in 100%) and the percentage
may be increased or decreased (e.g., 80 or 120 for 20% smaller or larger, respectively).

`nobox` prevents a "weighted box" from being drawn for each study; only markers for
point estimates are shown.

`nooverall` specifies that the overall estimate not be displayed, for example, when it is
inappropriate to meta-analyze across groups. (This option automatically enforces
the `nowt` option.)

`nowt` prevents the display of study weight on the graph.

`nostats` prevents the display of study statistics on the graph.

`counts` displays data counts (n/N) for each group when using binary data or the sample
size, mean, and standard deviation for each group if mean differences are used (the
latter is a new feature).

`group1(`*string*`)` and `group2(`*string*`)` may be used with the `counts` option, and the text
should contain the names of the two groups.

`effect(`*string*`)` allows the graph to name the summary statistic used when the effect
size and its standard error are declared.

`force` forces the x-axis scale to be in the range specified by `xlabel()`.

lcols(*varlist*) and rcols(*varlist*) define columns of additional data to the left or right of the graph. The first two columns on the right are automatically set to effect size and weight, unless suppressed by using the options nostats and nowt. If counts is used, this will be set as the third column. textsize() can be used to fine-tune the size of the text to achieve a satisfactory appearance. The columns are labeled with the variable label or the variable name if this is not defined. The first variable specified in lcols() is assumed to be the study identifier and this is used in the table output.

astext(#) specifies the percentage of the graph to be taken up by text. The default is 50%, and the percentage must be in the range 10–90.

double allows variables specified in lcols() and rcols() to run over two lines in the plot. This option may be of use if long strings are used.

nohet prevents the display of heterogeneity statistics in the graph.

summaryonly shows only summary estimates in the graph. This option may be of use for multiple subgroup analyses.

rfdist displays the confidence interval of the approximate predictive distribution of a future trial, based on the extent of heterogeneity. This option incorporates uncertainty in the location and spread of the random-effects distribution using the formula $t(\text{df}) \times \sqrt{\text{se2} + \text{tau2}}$, where t is the t distribution with $k - 2$ degrees of freedom, se2 is the squared standard error, and tau2 the heterogeneity statistic. The confidence interval is then displayed with lines extending from the diamond. With more than 3 studies, the distribution is inestimable and effectively infinite. It is thus displayed with dotted lines. Where heterogeneity is zero, there is still a slight extension as the t statistic is always greater than the corresponding normal deviate. For further information, see Higgins and Thompson (2006).

rflevel(#) specifies the coverage (e.g., 90%, 95%, 99%) for the confidence interval of the predictive distribution. The default is $S_level. See [U] **20.7 Specifying the width of confidence intervals**.

null(#) displays the null line at a user-defined value rather than at 0 or 1.

nulloff removes the null hypothesis line from the graph.

favours(*string* # *string*) applies a label saying something about the treatment effect to either side of the graph (strings are separated by the # symbol). This option replaces the feature available in b1title in the previous version of metan.

firststats(*string*) and secondstats(*string*) label overall user-defined estimates when these have been specified. Labels are displayed in the position usually given to the heterogeneity statistics.

boxopt(*marker_options*), diamopt(*line_options*), pointopt(*marker_options* | *marker_label_options*) ciopt(*line_options*), and olineopt(*line_options*) specify options for the graph routines within the program, allowing the user to alter the appearance of the graph. Any options associated with a particular graph command may be used, except some that would cause incorrect graph appearance. For example, diamonds are plotted using the `twoway pcspike` command, so options for line styles are available (see [G] *line_options*); however, altering the $x - -y$ orientation with the option `horizontal` or `vertical` is not allowed. So, `diamopt(lcolor(green) lwidth(thick))` feeds into a command such as `pcspike(y1 x1 y2 x2, lcolor(green) lwidth(thick))`.

boxopt(*marker_options*) controls the boxes and uses options for a weighted marker (e.g., shape and color, but not size). See [G] *marker_options*.

diamopt(*line_options*) controls the diamonds and uses options for `twoway pcspike` (not horizontal/vertical). See [G] *line_options*.

pointopt(*marker_options* | *marker_label_options*) controls the point estimate by using marker options. See [G] *marker_options* and [G] *marker_label_options*.

ciopt(*line_options*) controls the confidence intervals for studies by using options for `twoway pcspike` (not horizontal/vertical). See [G] *line_options*.

olineopt(*line_options*) controls the overall effect line with options for another line (not position). See [G] *line_options*.

classic specifies that solid black boxes without point estimate markers are used, as in the previous version of `metan`.

nowarning switches off the default display of a note warning that studies are weighted from random-effects analyses.

graph_options are any of the options documented in [G] *twoway_options*. These allow the addition of titles, subtitles, captions, etc.; control of margins, plot regions, graph size, and aspect ratio; and the use of schemes. Because titles may be added with *graph_options*, previous options such as `b2title` are no longer necessary.

11   Saved results from metan (macros)

As with many Stata commands, macros are left behind containing the results of the analysis. These include the pooled-effect size and its standard error; or, as described above regarding generated variables, the standard error of the log-effect size for odds and risk ratios. If two methods are specified, by using the option `second()`, some of these are repeated; for example, `r(ES)` and `r(ES_2)` give the pooled-effects estimates for each method. Subgroup statistics when using the `by()` option are not saved; if these are required for storage, it is recommended that a program be written that analyzes subgroups separately (perhaps using the `nograph` and `notable` options).

(Continued on next page)

Name	Second	Description
r(ES)	r(ES_2)	pooled-effect size (if the log option is specified with or or rr, this is the pooled log OR or log RR)
r(seES)	r(seES_2)	standard error of pooled-effect size with symmetrical CI, i.e., mean differences, risk difference, log OR, and log RR using log option
r(selogES)	r(selogES_2)	standard error of log OR or log RR when ORs or RRs are combined without the log option
r(ci_low)	r(ci_low_2)	lower CI of pooled-effect size
r(ci_upp)	r(ci_upp_2)	upper CI of pooled-effect size
r(z)		z-value of effect size
r(p_z)		p-value for significance of effect size
r(het)		chi-squared test for heterogeneity
r(df)		degrees of freedom (number of informative studies minus 1)
r(p_het)		p-value for significance of test for heterogeneity
r(i_sq)		the I^2 statistic
r(tau2)		estimated between-study variance (random-effects analyses only)
r(chi2)		chi-squared test for significance of odds ratio (fixed-effects OR only)
r(p_chi2)		p-value for the above test
r(rger)		overall event rate, group 1 (if binary data are combined)
r(cger)		overall event rate, group 2 (see above)
r(measure)		effect measure (e.g., RR, SMD)
r(method_1)	r(method_2)	analysis method (e.g., MH, DL)

Also, the following variables are added to the dataset by default (to override this use the `nokeep` option):

Variable name	Definition
_ES	Effect size (ES)
_seES	Standard error of ES
_LCI	Lower confidence limit for ES
_UCI	Upper confidence limit for ES
_WT	Study weight
_SS	Study sample size

12 Syntax for funnel

> **Editorial note:** The `funnel` command is no longer distributed with the `metan` package. More recent commands, `metafunnel` and `confunnel`, which use up-to-date Stata graphics and are described later in this collection, are recommended for funnel plots.

funnel [*precision_var effect_size*] [*if*] [*in*] [, *options*]

If the `funnel` command is invoked following `metan` with no parameters specified it will produce a standard funnel plot of precision (1/SE) against treatment effect. Addition of the `noinvert` option will produce a plot of standard error against treatment effect. The alternative sample size version of the funnel plot can be obtained by using the `sample` option (this automatically selects the `noinvert` option). Alternative plots can be created by specifying *precision_var* and *effect_size*. If the effect size is a relative risk or odds ratio, then the `xlog` graph option should be used to create a symmetrical plot.

13 Options for funnel

All options for `graph` are valid. Additionally, the following may be specified:

<u>sample</u> denotes that the y axis is the sample size and not a standard error.

<u>noinvert</u> prevents the values of the precision variable from being inverted.

<u>ysqrt</u> represents the y axis on a square-root scale.

<u>overall</u>(x) draws a dashed vertical line at the overall effect size given by x.

(Continued on next page)

14 Syntax for labbe

labbe *varlist* [*if*] [*in*] [*weight*] [, nowt <u>percent</u> or(*#*) rd(*#*) rr(*#*)
   rrn(*#*) null logit wgt(*weightvar*) symbol(*symbolstyle*) nolegend id(*idvar*)
   textsize(*#*) clockvar(*clockvar*) gap(*#*) *graph_options*]

15 Options for labbe

nowt declares that the plotted data points are to be the same size.

percent displays the event rates as percentages rather than proportions.

or(*#*) draws a line corresponding to a fixed odds ratio of *#*.

rd(*#*) draws a line corresponding to a fixed risk difference of *#*.

rr(*#*) draws a line corresponding to a fixed risk ratio of *#*. See also the rrn() option.

rrn(*#*) draws a line corresponding to a fixed risk ratio (for the nonevent) of *#*. The
   rr() and rrn() options may require explanation. Whereas the OR and RD are
   invariant to the definition of which of the binary outcomes is the "event" and which
   is the "nonevent", the RR is not. That is, while the command metan a b c d,
   or gives the same result as metan b a d c, or (with direction changed), an RR
   analysis does not. The L'Abbe plot allows the display of either or both to be
   superimposed risk difference.

null draws a line corresponding to a null effect (i.e., $p1 = p2$).

logit is for use with the or() option; it displays the probabilities on the logit scale,
   i.e., $\log(p/1-p)$. On the logit scale, the odds ratio is a linear effect, making it easier
   to assess the "fit" of the line.

wgt(*weightvar*) specifies alternative weighting by the specified variable; the default is
   sample size.

symbol(*symbolstyle*) allows the symbol to be changed (see [G] *symbolstyle*); the default
   being hollow circles (or points if weights are not used).

nolegend suppresses a legend from being displayed (the default if more than one line
   corresponding to effect measures are specified).

id(*idvar*) displays marker labels with the specified ID variable *idvar*. clockvar() and
   gap() may be used to fine-tune the display, which may become unreadable if studies
   are clustered together in the graph.

textsize(*#*) increases or decreases the text size of the ID label by specifying *#* to
   be more or less than unity. The default is usually satisfactory but may need to be
   adjusted.

clockvar(*clockvar*) specifies the position of *idvar* around the study point, as if it were a clock face (values must be integers; see [G] **clockposstyle**). This option may be used to organize labels where studies are clustered together. By default, labels are positioned to the left (9 o'clock) if above the null and to the right (3 o'clock) if below. Missing values in *clockvar* will be assigned the default position, so this need not be specified for all observations.

gap(*#*) increases or decreases the gap between the study marker and the ID label by specifying *#* to be more or less than unity. The default is usually satisfactory but may need to be adjusted.

graph_options specifies overall graph options that would appear at the end of a twoway graph command. This allows the addition of titles, subtitles, captions, etc., control of margins, plot regions, graph size, aspect ratio, and the use of schemes. See [G] **twoway_options** for a list of available options.

16 Example 1: Interventions in smoking cessation

Silagy and Ketteridge (1997)reported a systematic review of randomized controlled trials investigating the effects of physician advice on smoking cessation. In their review, they considered a meta-analysis of trials which have randomized individuals to receive either a minimal smoking cessation intervention from their family doctor or no intervention. An intervention was considered to be "minimal" if it consisted of advice provided by a physician during a single consultation lasting less than 20 minutes (possibly in combination with an information leaflet) with at most one follow-up visit. The outcome of interest was cessation of smoking. The data are presented below:

(Continued on next page)

```
. use example1

. describe

Contains data from example1.dta
  obs:            16
 vars:             6                              17 Jul 2008 13:28
 size:           320 (99.9% of memory free)
```

variable name	storage type	display format	value label	variable label
name	str8	%9s		
year	int	%8.0g		
a	byte	%8.0g		
r1	int	%8.0g		
c	byte	%8.0g		
r2	int	%8.0g		

```
Sorted by:

. list
```

	name	year	a	r1	c	r2
1.	Slama	1990	1	104	1	106
2.	Porter	1972	5	101	4	90
3.	Demers	1990	15	292	5	292
4.	Stewart	1982	11	504	4	187
5.	Page	1986	8	114	5	68
6.	Slama	1995	42	2199	5	929
7.	Haug	1994	20	154	7	109
8.	Russell	1979	34	1031	8	1107
9.	Wilson	1982	21	106	11	105
10.	McDowell	1985	12	85	11	78
11.	Janz	1987	28	144	12	106
12.	Wilson	1990	43	577	17	532
13.	Vetter	1990	34	237	20	234
14.	Higashi	1995	53	468	35	489
15.	Russell	1983	43	761	35	659
16.	Jamrozik	1984	77	512	58	549

We start by producing the data in the format of table 1, and pooling risk ratios by the Mantel–Haenszel fixed-effects method.

```
. generate b = r1-a

. generate d = r2-c

. label var name "Study"

. label var year "Year of publication"
```

```
. metan a b c d, rr lcols(name year) xlabel(0.1,0.2,0.5,1,2,5,10)
> title(Impact of physician advice in smoking cessation) boxsca(50)
        Study       |    RR     [95% Conf. Interval]     % Weight
--------------------+---------------------------------------------------
Slama               |  1.019     0.065     16.081           0.40
Porter              |  1.114     0.309      4.021           1.71
Demers              |  3.000     1.105      8.147           2.02
Stewart             |  1.020     0.329      3.165           2.36
Page                |  0.954     0.325      2.800           2.53
Slama               |  3.549     1.409      8.941           2.84
Haug                |  2.022     0.886      4.615           3.32
Russell             |  4.563     2.122      9.811           3.12
Wilson              |  1.891     0.960      3.724           4.47
McDowell            |  1.001     0.469      2.137           4.64
Janz                |  1.718     0.917      3.219           5.59
Wilson              |  2.332     1.347      4.038           7.16
Vetter              |  1.678     0.996      2.829           8.14
Higashi             |  1.582     1.053      2.379          13.85
Russell             |  1.064     0.689      1.642          15.18
Jamrozik            |  1.424     1.035      1.958          22.65
--------------------+---------------------------------------------------
M-H pooled RR       |  1.676     1.440      1.951         100.00
--------------------+---------------------------------------------------

    Heterogeneity chi-squared =  21.51 (d.f. = 15) p = 0.121
    I-squared (variation in RR attributable to heterogeneity) =  30.3%

    Test of RR=1 : z=   6.66 p = 0.000
```

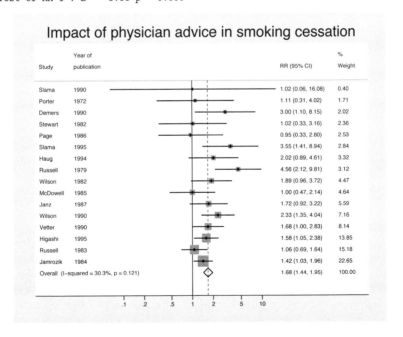

Figure 1. Forest plot for example 1

It appears that there is a significant benefit of such minimal intervention. The non-significance of the test for heterogeneity suggests that the differences between the studies are explicable by random variation, although this test has low statistical power. The L'Abbé plot provides an alternative way of displaying the data which allows inspection of the variability in experimental and control group event rates.

```
. labbe a b c d, xlabel(0,0.1,0.2,0.3) ylabel(0,0.1,0.2,0.3) psize(50)
> t1(Impact of physician advice in smoking cessation:)
> t2(Proportion of patients ceasing to smoke)
> l1(Physician intervention group patients) b2(Control group patients)
(See figure 2 below)
```

A funnel plot can be used to investigate the possibility that the studies which were included in the review were a biased selection. The alternative command `metabias` (Steichen 1998) additionally gives a formal test for nonrandom inclusion of studies in the review.

```
. funnel, xlog ylabel(0,2,4,6) xlabel(0.5,1,2,5) xli(1) overall(1.68)
> b2(Risk Ratio)
(See figure 3 below)
```

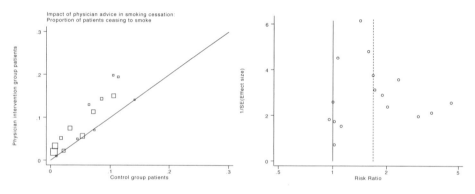

Figure 2. L'Abbé plot for example 1   Figure 3. Funnel plot for example 1

Interpretation of funnel plots can be difficult, as a certain degree of asymmetry is to be expected by chance.

17   Example 2

D'Agostino and Weintraub (1995) reported a meta-analysis of the effects of antihistamines in common cold preparations on the severity of sneezing and runny nose. They combined data from nine randomized trials in which participants with new colds were randomly assigned to an active antihistamine treatment or placebo. The effect of the treatment was measured as the change in severity of runny nose following one day's treatment. The trials used a variety of scales for measuring severity. Due to this, standardized mean differences are used in the analysis. We choose to use Cohen's method (the default) to compute the standardized mean difference.

```
. use example2, clear
. list
```

	n1	mean1	sd1	n2	mean2	sd2
1.	11	.273	.786	16	-.188	.834
2.	128	.932	.593	136	.81	.556
3.	63	.73	.745	64	.578	.773
4.	22	.35	1.139	22	.339	.744
5.	16	.422	2.209	15	-.17	1.374
6.	39	.256	1.666	41	.537	1.614
7.	21	2.831	1.753	21	1.396	1.285
8.	13	2.687	1.607	8	1.625	2.089
9.	194	.49	.895	193	.264	.828

```
. metan n1 mean1 sd1 n2 mean2 sd2, xlabel(-1.5,-1,-0.5,0,0.5,1,1.5)
> title("Effect of antihistamines on cold severity")
> xtitle("Standardized mean difference") textsize(125)
              Study    |    SMD   [95% Conf. Interval]    % Weight
--------------------+-----------------------------------------------
1                     |   0.566   -0.218     1.349          2.48
2                     |   0.212   -0.030     0.455         26.01
3                     |   0.200   -0.149     0.549         12.53
4                     |   0.011   -0.580     0.602          4.36
5                     |   0.319   -0.390     1.029          3.03
6                     |  -0.171   -0.611     0.268          7.90
7                     |   0.934    0.295     1.572          3.74
8                     |   0.590   -0.310     1.491          1.88
9                     |   0.262    0.062     0.462         38.06
--------------------+-----------------------------------------------
I-V pooled SMD        |   0.237    0.113     0.360        100.00
--------------------+-----------------------------------------------

  Heterogeneity chi-squared =   9.91 (d.f. = 8) p = 0.271
  I-squared (variation in SMD attributable to heterogeneity) =  19.3%

  Test of SMD=0 : z=   3.76 p = 0.000
```

(Continued on next page)

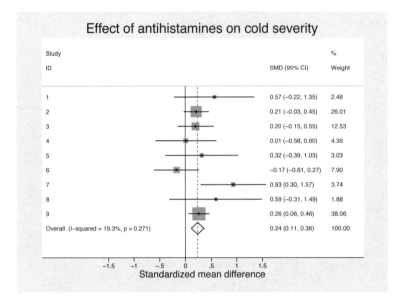

Figure 4. Forest plot for example 2

The patients given antihistamines appear to have a greater reduction in severity of cold symptoms in the first 24 hours of treatment. Again the between-study differences are explicable by random variation.

18   Formulas

19   Individual study responses: binary outcomes

For study i denote the cell counts as in table 1, and let $n_{1i} = a_i + b_i$, $n_{2i} = c_i + d_i$ (the number of participants in the treatment and control groups respectively) and $N_i = n_{1i} + n_{2i}$ (the number in the study). For the Peto method, the individual odds ratios are given by

$$\widehat{\mathrm{OR}}_i = \exp\left[\{a_i - E\left(a_i\right)\}/v_i\right]$$

with its logarithm having standard error

$$\mathrm{se}\{\ln(\widehat{\mathrm{OR}}_i)\} = \sqrt{1/v_i}$$

where $E(a_i) = n_{1i}(a_i + c_i)/N_i$ (the expected number of events in the exposure group) and

$$v_i = \{n_{1i}n_{2i}(a_i + c_i)(b_i + d_i)\}/\{N_i^2(N_i - 1)\} \text{ (the hypergeometric variance of } a_i\text{)}.$$

For other methods of combining trials, the odds ratio for each study is given by

$$\widehat{OR}_i = a_i d_i / b_i c_i$$

the standard error of the log odds-ratio being

$$se\{\ln(\widehat{OR}_i)\} = \sqrt{1/a_i + 1/b_i + 1/c_i + 1/d_i}$$

The risk ratio for each study is given by

$$\widehat{RR}_i = (a_i/n_{1i})/(c_i/n_{2i})$$

the standard error of the log risk-ratio being

$$se\{\ln(\widehat{RR}_i)\} = \sqrt{1/a_i + 1/c_i - 1/n_{1i} - 1/n_{2i}}$$

The risk difference for each study is given by

$$\widehat{RD}_i = (a_i/n_{1i}) - (c_i/n_{2i}) \text{ with standard error } se(\widehat{RD}_i) = \sqrt{a_i b_i / n_{1i}^3 + c_i d_i / n_{2i}^3}$$

where zero cells cause problems with computation of the standard errors, 0.5 is added to all cells (a_i, b_i, c_i, d_i) for that study.

20   Individual study responses: continuous outcomes

Denote the number of subjects, mean, and standard deviation as in table 1, and let

$$N_i = n_{1i} + n_{2i}$$

and

$$s_i = \sqrt{\{(n_{1i} - 1)sd_{1i}^2 + (n_{2i} - 1)sd_{2i}^2\}/(N_i - 2)}$$

be the pooled standard deviation of the two groups. The weighted mean difference is given by

$$\widehat{WMD}_i = m_{1i} - m_{2i} \text{ with standard error } se(\widehat{WMD}_i) = \sqrt{sd_{1i}^2/n_{1i} + sd_{2i}^2/n_{2i}}$$

There are three formulations of the standardized mean difference. The default is the measure suggested by Cohen (Cohen's d), which is the ratio of the mean difference to the pooled standard deviation s_i; i.e.,

$$\widehat{d}_i = (m_{1i} - m_{2i})/s_i \text{ with standard error } se(\widehat{d}_i) = \sqrt{N_i/(n_{1i}n_{2i}) + \widehat{d}_i^2/2(N_i - 2)}$$

Hedges suggested a small-sample adjustment to the mean difference (Hedges adjusted g), to give

$$\widehat{g}_i = \{(m_{1i} - m_{2i})/s_i\}\{1 - 3/(4N_i - 9)\} \text{ with standard error}$$
$$se(\widehat{g}_i) = \sqrt{N_i/(n_{1i}n_{2i}) + \widehat{g}_i^2/2(N_i - 3.94)}$$

Glass suggested using the control group standard deviation as the best estimate of the scaling factor to give the summary measure (Glass's $\widehat{\Delta}$), where

$$\widehat{\Delta}_i = (m_{1i} - m_{2i})/sd_{2i}, \text{ with standard error } se(\Delta_i) = \sqrt{N_i/(n_{1i}n_{2i}) + \widehat{\Delta}_i^2/2(n_{2i} - 1)}$$

21   Mantel–Haenszel methods for combining trials

For each study, the effect size from each trial $\hat{\theta}_i$ is given weight w_i in the analysis. The overall estimate of the pooled effect, $\hat{\theta}_{MH}$ is given by

$$\hat{\theta}_{MH} = \left(\sum w_i \hat{\theta}_i\right)/\left(\sum w_i\right)$$

For combining odds ratios, each study's OR is given weight

$$w_i = b_i c_i / N_i$$

and the logarithm of $\widehat{OR}_{MH}$ has standard error given by

$$se\{\ln(\widehat{OR}_{MH})\} = \sqrt{(PR)/2R^2 + \{(PS + QR)/2(R \times S)\} + (QS)/2S^2}$$

where

$$R = \sum a_i d_i / N_i \qquad S = \sum b_i c_i / N_i$$

$$PR = \sum (a_i + d_i) a_i d_i / N_i^2 \qquad PS = \sum (a_i + d_i) b_i c_i / N_i^2$$

$$QR = \sum (b_i + c_i) a_i d_i / N_i^2 \qquad QS = \sum (b_i + c_i) b_i c_i / N_i^2$$

For combining risk ratios, each study's RR is given weight

$$w_i = (n_{1i} c_i) / N_i$$

and the logarithm of $\widehat{RR}_{MH}$ has standard error given by

$$se\{\ln(\widehat{RR}_{MH})\} = \sqrt{P/(R \times S)}$$

where

$$P = \sum (n_{1i} n_{2i} (a_i + c_i) - a_i c_i N_i)/N_i^2 \qquad R = \sum a_i n_{2i} / N_i \qquad S = \sum c_i n_{1i} / N_i$$

For risk differences, each study's RD has the weight

$$w_i = n_{1i} n_{2i} / N_i$$

and $\widehat{RD}_{MH}$ has standard error given by

$$se\{\widehat{RD}_{MH}\} = \sqrt{(P/Q^2)}$$

where

$$P = \sum (a_i b_i n_{2i}^3 + c_i d_i n_{1i}^3)/n_{1i} n_{2i} N_i^2; \qquad Q = \sum n_{1i} n_{2i} / N_i$$

The heterogeneity statistic is given by

$$Q = \sum w_i (\hat{\theta}_i - \hat{\theta}_{MH})^2$$

where θ is the log odds-ratio, log relative-risk, or risk difference. Under the null hypothesis that there are no differences in treatment effect between trials, this follows a χ^2 distribution on $k - 1$ degrees of freedom.

22 Inverse variance methods for combining trials

Here, when considering odds ratios or risk ratios, we define the effect size θ_i to be the natural logarithm of the trial's OR or RR; otherwise, we consider the summary statistic (RD, SMD, or WMD) itself. The individual effect sizes are weighted according to the reciprocal of their variance (calculated as the square of the standard errors given in the individual study section above) giving

$$w_i = 1/\mathrm{se}(\widehat{\theta}_i)^2$$

These are combined to give a pooled estimate

$$\widehat{\theta}_{\mathrm{IV}} = (\textstyle\sum w_i \widehat{\theta}_i)/(\sum w_i)$$

with

$$\mathrm{se}\{\widehat{\theta}_{\mathrm{IV}}\} = 1/\sqrt{\textstyle\sum w_i}$$

The heterogeneity statistic is given by a similar formula as for the Mantel–Haenszel method, using the inverse variance form of the weights, w_i

$$Q = \textstyle\sum w_i(\widehat{\theta}_i - \widehat{\theta}_{\mathrm{IV}})^2$$

23 Peto's assumption free method for combining trials

Here, the overall odds ratio is given by

$$\widehat{\mathrm{OR}}_{\mathrm{Peto}} = \exp\{\textstyle\sum w_i \ln(\widehat{\mathrm{OR}}_i)/\sum w_i\}$$

where the odds ratio $\widehat{\mathrm{OR}}_i$ is calculated using the approximate method described in the individual trial section, and the weights, w_i are equal to the hypergeometric variances, v_i.

The logarithm of the odds ratio has standard error

$$\mathrm{se}\{\ln(\widehat{\mathrm{OR}}_{\mathrm{Peto}})\} = 1/\sqrt{\textstyle\sum w_i}$$

The heterogeneity statistic is given by

$$Q = \textstyle\sum w_i\{(\ln \widehat{\mathrm{OR}}_i)^2 - (\ln \widehat{\mathrm{OR}}_{\mathrm{Peto}})^2\}$$

24 DerSimonian and Laird random-effects models

Under the random-effects model, the assumption of a common treatment effect is relaxed, and the effect sizes are assumed to have a distribution

$$\theta_i \sim N(\theta, \tau^2)$$

The estimate of τ^2 is given by

$$\widehat{\tau}^2 = \max[\{Q - (k-1)\}/\{\textstyle\sum w_i - (\sum(w_i^2)/\sum w_i)\}, 0]$$

The estimate of the combined effect for heterogeneity may be taken as either the Mantel–Haenszel or the inverse variance estimate. Again, for odds ratios and risk ratios, the effect size is taken as the natural logarithm of the OR and RR. Each study's effect size is given weight

$$w_i = 1/\{\text{se}(\widehat{\theta}_i)^2 + \widehat{\tau}^2\}$$

The pooled effect size is given by

$$\widehat{\theta}_{\text{DL}} = (\textstyle\sum w_i \widehat{\theta}_i)/(\textstyle\sum w_i)$$

and

$$\text{se}\{\widehat{\theta}_{\text{DL}}\} = 1/\sqrt{\textstyle\sum w_i}$$

Note that in the case where the heterogeneity statistic Q is less than or equal to its degrees of freedom $(k - 1)$, the estimate of the between trial variation, $\widehat{\tau}^2$, is zero, and the weights reduce to those given by the inverse variance method.

25   Confidence intervals

The $100(1 - \alpha)\%$ confidence interval for $\widehat{\theta}$ is given by

$$\widehat{\theta} - \text{se}(\widehat{\theta})\Phi(1 - \alpha/2), \quad \text{to} \quad \widehat{\theta} + \text{se}(\widehat{\theta})\Phi(1 - \alpha/2)$$

where $\widehat{\theta}$ is the log odds-ratio, log relative-risk, risk difference, mean difference, or standardized mean difference, and Φ is the standard normal distribution function. The Cornfield confidence intervals for odds ratios are calculated as explained in the Stata manual for the `epitab` command.

26   Test statistics

In all cases, the test statistic is given by

$$z = \widehat{\theta}/\text{se}(\widehat{\theta})$$

where the odds ratio or risk ratio is again considered on the log scale.

For odds ratios pooled by method of Mantel and Haenszel or Peto, an alternative test statistic is available, which is the χ^2 test of the observed and expected events rate in the exposure group. The expectation and the variance of a_i are as given earlier in the Peto odds-ratio section. The test statistic is

$$\chi^2 = [\textstyle\sum\{a_i - E(a_i)\}]^2/\textstyle\sum \text{Var}(a_i)$$

on one degree of freedom. Note that in the case of odds ratios pooled by method of Peto, the two test statistics are identical; the χ^2 test statistic is simply the square of the z score.

27   Acknowledgments

The statistical methods programmed in `metan` utilize several of the algorithms used by the MetaView software (part of the Cochrane Library), which was developed by Gordon Dooley of Update Software, Oxford and Jonathan Deeks of the Statistical Methods Working Group of the Cochrane Collaboration. We have also used a subroutine written by Patrick Royston of the Royal Postgraduate Medical School, London.

28   References

Breslow, N. E., and N. E. Day. 1993. *Statistical Methods in Cancer Research: Volume I—The Analysis of Case–Control Studies.* Lyon, UK: International Agency for Research on Cancer.

D'Agostino, R. B., and M. Weintraub. 1995. Meta-analysis: A method for synthesizing research. *Clinical Pharmacology and Therapeutics* 58: 605–616.

DerSimonian, R., and N. Laird. 1986. Meta-analysis in clinical trials. *Controlled Clinical Trials* 7: 177–188.

Egger, M., G. Davey Smith, M. Schneider, and C. Minder. 1997. Bias in meta-analysis detected by a simple, graphical test. *British Medical Journal* 315: 629–634.

Fleiss, J. L. 1993. The statistical basis of meta-analysis. *Statistical Methods in Medical Research* 2: 121–145.

Glass, G. V., B. McGaw, and M. L. Smith. 1981. *Meta-Analysis in Social Research.* Beverly Hills, CA: Sage.

Greenland, S., and J. Robins. 1985. Estimation of a common effect parameter from sparse follow-up data. *Biometrics* 41: 55–68.

Greenland, S., and A. Salvan. 1990. Bias in the one-step method for pooling study results. *Statistics in Medicine* 9: 247–252.

Hedges, L. V., and I. Olkin. 1985. *Statistical Methods for Meta-analysis.* San Diego, CA: Academic Press.

L'Abbé, K. A., A. S. Detsky, and K. O'Rourke. 1987. Meta-analysis in clinical research. *Annals of Internal Medicine* 107: 224–233.

Mantel, N., and W. Haenszel. 1959. Statistical aspects of the analysis of data from retrospective studies of diseases. *Journal of the National Cancer Institute* 22: 719–748.

Robins, J., S. Greenland, and N. E. Breslow. 1986. A general estimator for the variance of the Mantel–Haenszel odds ratio. *American Journal of Epidemiology* 124: 719–723.

Rosenthal, R. 1994. Parametric measures of effect size. In *The Handbook of Research Synthesis*, ed. H. Cooper and L. V. Hedges. New York: Russell Sage Foundation.

Silagy, C., and S. Ketteridge. 1997. Physician advice for smoking cessation. In *Tobacco Addiction Module of the Cochrane Database of Systematic Reviews*, ed. T. Lancaster, C. Silagy, and D. Fullerton. Oxford: The Cochrane Collaboration. Available in the Cochrane Library (subscription database and CDROM), issue 4.

Sinclair, J. C., and M. B. Bracken. 1992. *Effective Care of the Newborn Infant*. Oxford: Oxford University Press.

Steichen, T. J. 1998. sbe19: Tests for publication bias in meta-analysis. *Stata Technical Bulletin* 41: 9–15. Reprinted in *Stata Technical Bulletin Reprints*, vol. 7, pp. 125–133. College Station, TX: Stata Press. (Reprinted in this collection on pp. 151–161.)

Yusuf, S., R. Peto, J. Lewis, R. Collins, and P. Sleight. 1985. Beta blockade during and after myocardial infarction: An overview of the randomized trials. *Progress in Cardiovascular Diseases* 27: 335–371.

The Stata Journal (2008)
8, Number 1, pp. 3–28

metan: fixed- and random-effects meta-analysis

Ross J. Harris
Centre for Infections
Health Protection Agency
London, UK
ross.harris@hpa.org.uk

Michael J. Bradburn
Clinical Trials Research Unit
ScHARR
Sheffield, UK
m.bradburn@sheffield.ac.uk

Jonathan J. Deeks
Unit of Public Health, Epidemiology, and Biostatistics
University of Birmingham
Birmingham, UK
j.deeks@bham.ac.uk

Roger M. Harbord
Department of Social Medicine
University of Bristol
Bristol, UK

Douglas G. Altman
Centre for Statistics in Medicine
University of Oxford
Oxford, UK

Jonathan A. C. Sterne
Department of Social Medicine
University of Bristol
Bristol, UK

Abstract. This article describes updates of the meta-analysis command `metan` and options that have been added since the command's original publication (Bradburn, Deeks, and Altman, metan – an alternative meta-analysis command, *Stata Technical Bulletin Reprints*, vol. 8, pp. 86–100). These include version 9 graphics with flexible display options, the ability to meta-analyze precalculated effect estimates, and the ability to analyze subgroups by using the `by()` option. Changes to the output, saved variables, and saved results are also described.

Keywords: sbe24_2, metan, meta-analysis, forest plot

1 Introduction

Meta-analysis is a two-stage process involving the estimation of an appropriate summary statistic for each of a set of studies followed by the calculation of a weighted average of these statistics across the studies (Deeks, Altman, and Bradburn 2001). Odds ratios, risk ratios, and risk differences may be calculated from binary data, or a difference in means obtained from continuous data. Alternatively, precalculated effect estimates and their standard errors from each study may be pooled, for example, adjusted log odds-ratios from observational studies. The summary statistics from each study can be combined by using a variety of meta-analytic methods, which are classified as fixed-effects models in which studies are weighted according to the amount of information they contain; or random-effects models, which incorporate an estimate of between-study variation (heterogeneity) in the weighting. A meta-analysis will customarily include a forest plot, in which results from each study are displayed as a square and a horizontal line, representing the intervention effect estimate together with its confidence interval. The area of the square reflects the weight that the study contributes to the meta-analysis. The combined-effect estimate and its confidence interval are represented by a diamond.

Here we present updates to the `metan` command and other previously undocumented additions that have been made since its original publication (Bradburn, Deeks, and Altman 1998). New features include

- Version 9 graphics

- Flexible display of tabular data in the forest plot

- Results from a second type of meta-analysis displayed in the same forest plot

- `by()` group processing

- Analysis of precalculated effect estimates

- Prediction intervals for the intervention effect in a new study from random-effects analyses

There are a substantial number of options for the `metan` command because of the variety of meta-analytic techniques and the need for flexible graphical displays. We recommend that new users not try to learn everything at once but to learn the basics and build from there as required. Clickable examples of `metan` are available in the help file, and the dialog box may also be a good way to start using `metan`.

2   Example data

The dataset used in subsequent examples is taken from the meta-analysis published as table 1 in Colditz et al. (1994, 699). The aim of the analysis was to quantify the efficacy of BCG vaccine against tuberculosis, and data from 11 trials are included here. There was considerable between-trial heterogeneity in the effect of the vaccine; it has been suggested that this might be explained by the latitude of the region in which the trial was conducted (Fine 1995).

▷ **Example**

Details of the dataset are shown below by using `describe` and `list` commands.

```
. use bcgtrial
(BCG and tuberculosis)

. describe

Contains data from bcgtrial.dta
  obs:            11                          BCG and tuberculosis
  vars:           12                          31 May 2007 17:11
  size:           693 (99.9% of memory free)  (_dta has notes)
```

variable name	storage type	display format	value label	variable label
trial	byte	%8.0g		Trial number
trialnam	str14	%14s		Trial name
authors	str20	%20s		Authors of trial
startyr	int	%8.0g		Year trial started
latitude	byte	%8.0g		Latitude of trial area
alloc	byte	%33.0g	alloc	Allocation method
tcases	int	%8.0g		BCG vaccinated cases
tnoncases	float	%9.0g		BCG vaccinated noncases
ccases	int	%8.0g		Unvaccinated cases
cnoncases	float	%9.0g		Unvaccinated noncases
ttotal	long	%12.0g		BCG vaccinated population
ctotal	long	%12.0g		Unvaccinated population

```
Sorted by: startyr  authors

. list trialnam startyr tcases tnoncases ccases cnoncases, clean noobs
> abbreviate(10)
```

trialnam	startyr	tcases	tnoncases	ccases	cnoncases
Canada	1933	6	300	29	274
Northern USA	1935	4	119	11	128
Chicago	1941	17	1699	65	1600
Georgia (Sch)	1947	5	2493	3	2338
Puerto Rico	1949	186	50448	141	27197
Georgia (Comm)	1950	27	16886	29	17825
Madanapalle	1950	33	5036	47	5761
UK	1950	62	13536	248	12619
South Africa	1965	29	7470	45	7232
Haiti	1965	8	2537	10	619
Madras	1968	505	87886	499	87892

Trial name and number identify each study, and we have information on the authors and the year the trial started. There are also two variables relating to study characteristics: the latitude of the area in which the trial was carried out, and the method of allocating patients to the vaccine and control groups—either at random or in some systematic way. The variables tcases, tnoncases, ccases, and cnoncases contain the data from the 2×2 table from each study (the number of cases and noncases in the vaccination group and nonvaccination group). The variables ttotal and ctotal are the total number of individuals (the sum of the cases and noncases) in the vaccine and control groups. Displayed below is the 2×2 table for the first study (Canada, 1933):

	cases	noncases	total
treated	6	300	306
control	29	274	303

The risk ratio (RR), log risk-ratio (log RR), standard error of log RR (SE log RR), 95% confidence interval (CI) for log RR, and 95% CI for RR may be calculated as follows (see, for example, Kirkwood and Sterne 2001).

$$\text{Risk in treated population} = \frac{\texttt{tcases}}{\texttt{ttotal}} = \frac{6}{306} = 0.0196$$

$$\text{Risk in control population} = \frac{\texttt{ccases}}{\texttt{ctotal}} = \frac{29}{303} = 0.0957$$

$$\text{RR} = \frac{\text{Risk in treated population}}{\text{Risk in control population}} = \frac{0.0196}{0.0957} = 0.2049$$

$$\log \text{RR} = \log(\text{RR}) = -1.585$$

$$\text{SE}(\log \text{RR}) = \sqrt{\frac{1}{\texttt{tcases}} + \frac{1}{\texttt{ccases}} - \frac{1}{\texttt{ttotal}} - \frac{1}{\texttt{ctotal}}}$$

$$= \sqrt{\frac{1}{6} + \frac{1}{29} - \frac{1}{306} - \frac{1}{303}} = 0.441$$

$$95\% \text{ CI for } \log \text{RR} = \log \text{RR} \pm 1.96 \times \text{SE}(\log \text{RR}) = -2.450 \text{ to } -0.720$$

$$95\% \text{ CI for RR} = \exp(-2.450) \text{ to } \exp(-0.720) = 0.086 \text{ to } 0.486$$

◁

3   Syntax

metan *varlist* [*if*] [*in*] [,

    [*binary_data_options* | *continuous_data_options* | *precalculated_effect_estimates_options*]
    measure_and_model_options   output_options   forest_plot_options]

binary_data_options

    or rr rd fixed random fixedi randomi peto cornfield chi2 breslow
    <u>nointeger</u> cc(#)

continuous_data_options

    cohen hedges glass nostandard fixed random <u>nointeger</u>

precalculated_effect_estimates_options

    `fixed random`

measure_and_model_options

    `wgt(`*wgtvar*`) second(`*model* | *estimates_and_description*`)`
    `first(`*estimates_and_description*`)`

output_options

    `by(`*byvar*`) nosubgroup sgweight log eform efficacy i̲level(#)`
    `o̲level(#) sortby(`*varlist*`)`
    `label([namevar = `*namevar*`], [yearvar = `*yearvar*`]) nokeep notable`
    `nograph nosecsub`

forest_plot_options

    `xl̲abel(#, ...) xt̲ick(#, ...) boxsca(#) textsize(#) nobox nooverall`
    `nowt nostats counts group1(`*string*`) group2(`*string*`) effect(`*string*`) force`
    `lcols(`*varlist*`) rcols(`*varlist*`) astext(#) double nohet summaryonly rfdist`
    `rfl̲evel(#) null(#) nulloff favours(`*string* # *string*`) firststats(`*string*`)`
    `secondstats(`*string*`) boxopt(`*marker_options*`) diamopt(`*line_options*`)`
    `pointopt(`*marker_options* | *marker_label_options*`) ciopt(`*line_options*`)`
    `olineopt(`*line_options*`) classic nowarning` *graph_options*

For a full description of the syntax, see Bradburn, Deeks, and Altman (1998). We will focus on the new options, most of which come under *forest_plot_options*; previously undocumented options such as `by()` (and related options), `breslow`, `cc()`, `nointeger`; and changes to the output such as the display of the I^2 statistic. Syntax will be explained in the appropriate sections.

4 Basic use

4.1 2×2 data

For binary data, the input variables required by `metan` should contain the cells of the 2×2 table; i.e., the number of individuals who did and did not experience the outcome event in the treatment and control groups for each study. When analyzing 2×2 data a range of methods are available. The default is the Mantel–Haenszel method (`fixed`). The inverse-variance fixed-effects method (`fixedi`) or the Peto method for estimating summary odds ratios (`peto`) may also be chosen. The DerSimonian and Laird random-effects method may be specified with `random`. See Deeks, Altman, and Bradburn (2001) for a discussion of these methods.

4.2   Display options

Previous versions of the `metan` command used the syntax `label(namevar = ` *namevar* `,` `yearvar = ` *yearvar*`)` to specify study information in the table and forest plot. This syntax still functions but has been superseded by the more flexible `lcols(`*varlist*`)` and `rcols(`*varlist*`)` options. The use of these options is described in more detail in section 5. The option `favours(`*string* `#` *string*`)` allows the user to display text information about the direction of the treatment effect, which appears under the graph (e.g., exposure good, exposure bad). `favours()` replaces the option `b2title()`. The `#` is required to split the two strings, which appear to either side of the null line.

▷ **Example**

Here we use `metan` to derive an inverse-variance weighted (fixed effect) meta-analysis of the BCG trial data. Risk ratios are specified as the summary statistic, and the trial name and the year the trial started are displayed in the forest plot using `lcols()` (see section 5).

```
. metan tcases tnoncases ccases cnoncases, rr fixedi lcols(trialnam startyr)
> xlabel(0.1, 10) favours(BCG reduces risk of TB # BCG increases risk of TB)
```

Study	RR	[95% Conf. Interval]		% Weight
Canada	0.205	0.086	0.486	1.11
Northern USA	0.411	0.134	1.257	0.66
Chicago	0.254	0.149	0.431	2.96
Georgia (Sch)	1.562	0.374	6.528	0.41
Puerto Rico	0.712	0.573	0.886	17.42
Georgia (Comm)	0.983	0.582	1.659	3.03
Madanapalle	0.804	0.516	1.254	4.22
UK	0.237	0.179	0.312	10.81
South Africa	0.625	0.393	0.996	3.83
Haiti	0.198	0.078	0.499	0.97
Madras	1.012	0.895	1.145	54.58
I-V pooled RR	0.730	0.667	0.800	100.00

```
Heterogeneity chi-squared = 125.63 (d.f. = 10) p = 0.000
I-squared (variation in RR attributable to heterogeneity) =  92.0%

Test of RR=1 : z=   6.75 p = 0.000
```

The output table contains effect estimates (here RRs), CIs, and weights for each study, followed by the overall (combined) effect estimate. The results for the Canada study are identical to those derived in section 2. Heterogeneity statistics relating to the extent that RRs vary between studies are displayed, including the I^2 statistic, which is a previously undocumented addition. The I^2 statistic (see section 9.1) is the percentage of between-study heterogeneity that is attributable to variability in the true treatment effect, rather than sampling variation (Higgins and Thompson 2004, Higgins et al. 2003). Here there is substantial between-study heterogeneity. Finally, a test of the null hypothesis that the vaccine has no effect (RR=1) is displayed. There is strong evidence against the null hypothesis, but the presence of between-study heterogeneity means that the fixed-effects assumption (that the true treatment effect is the same in each study) is incorrect. The forest plot displayed by the command is shown in figure 1.

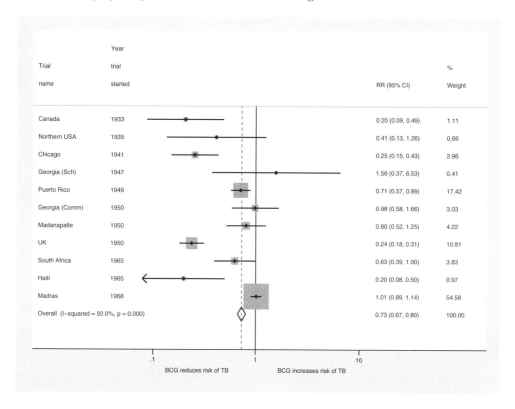

Figure 1. Forest plot displaying an inverse-variance weighted fixed-effects meta-analysis of the effect of BCG vaccine on incidence of tuberculosis.

◁

4.3   Precalculated effect estimates

The `metan` command may also be used to meta-analyze precalculated effect estimates, such as log odds-ratios and their standard errors or 95% CI, using syntax similar to the alternative Stata meta-analysis command `meta` (Sharp and Sterne 1997, 1998). Here only the inverse-variance fixed-effects and DerSimonian and Laird random-effects methods are available, because other methods require the 2×2 cell counts or the means and standard deviations in each group. The `fixed` option produces an inverse-variance weighted analysis when precalculated effect estimates are analyzed.

When analyzing ratio measures (RRs or odds ratios), the log ratio with its standard error or 95% CI should be used as inputs to the command. The `eform` option can then be used to display the output on the ratio scale (as for the `meta` command).

▷ **Example**

We will illustrate this feature by generating the log RR and its standard error in each study from the 2×2 data, and then by meta-analyzing these variables.

```
. gen logRR = ln( (tcases/ttotal) / (ccases/ctotal) )
. gen selogRR = sqrt( 1/tcases +1/ccases -1/ttotal -1/ctotal )
. metan logRR selogRR, fixed eform nograph
```

Study	ES	[95% Conf. Interval]		% Weight
(table of study results omitted)				
I-V pooled ES	0.730	0.667	0.800	100.00

```
Heterogeneity chi-squared = 125.63 (d.f. = 10) p = 0.000
I-squared (variation in ES attributable to heterogeneity) =  92.0%

Test of ES=1 : z=   6.75 p = 0.000
```

The results are identical to those derived directly from the 2×2 data in section 4.1; we would have observed minor differences if the default Mantel–Haenszel method had been used previously. When analyzing precalculated estimates, `metan` does not know what these measures are, so the summary estimate is named "ES" (effect size) in the output.

◁

4.4 Specifying two analyses

`metan` now allows the display of a second meta-analytic estimate in the same output table and forest plot. A typical use is to compare fixed-effects and random-effects analyses, which can reveal the presence of small-study effects. These may result from publication or other biases (Sterne, Gavaghan, and Egger 2000). See Poole and Greenland (1999) for a discussion of the ways in which fixed-effects and random-effects analyses may differ. The syntax is to specify the method for the second meta-analytic estimate as second(*method*), where *method* is any of the standard `metan` options.

▷ **Example**

Here we use `metan` to analyze 2×2 data as in section 4.1, specifying an inverse-variance weighted (fixed effect) model for the first method and a DerSimonian and Laird (random effects) model for the second method:

```
. metan tcases tnoncases ccases cnoncases, rr fixedi second(random)
> lcols(trialnam startyr) nograph
              Study       |    RR   [95% Conf. Interval]   % Weight
```

(table of study results omitted)

```
I-V pooled RR            |   0.730   0.667     0.800       100.00
D+L pooled RR            |   0.508   0.336     0.769       100.00
```

```
   Heterogeneity chi-squared = 125.63 (d.f. = 10) p = 0.000
   I-squared (variation in RR attributable to heterogeneity) =  92.0%

   Test of RR=1 : z=   6.75 p = 0.000
```

The results of the second analysis are displayed in the table: a forest plot using the `second()` option is derived in the next section and displayed in figure 2. The protective effect of BCG against tuberculosis appears greater in the random-effects analysis than in the fixed-effects analysis, although CI is wider. This reflects the greater uncertainty in the random-effects analysis, which allows for the true effect of the vaccine to vary between studies. Random-effects analyses give relatively greater weight to smaller studies than fixed-effects analyses, and so these results suggest that the estimated effect of BCG was greater in the smaller studies. It is also possible to supply a precalculated pooled-effect estimate with `second()`; see section 7.2 for details.

◁

5   Displaying data columns in graphs

The options `lcols(`*varlist*`)` and `rcols(`*varlist*`)` produce columns to the left or right of the forest plot. String (character) or numeric variables can be displayed. If numeric variables have value labels, these will be displayed in the graph. If the variable itself is labeled, this will be used as the column header, allowing meaningful names to be used. Up to four lines are used for the heading, so names can be long without taking up too much graph width.

The first variable in `lcols()` is used to identify studies in the table output, and summary statistics and study weight are always the first columns on the right of the forest plot. These can be switched off by using the options `nostats` and `nowt`, but the order cannot be changed.

If lengthy string variables are to be displayed, the `double` option may be used to allow output to spread over two lines per study in the forest plot. The percentage of the forest plot given to text may be adjusted using `astext(`*#*`)`, which can be between 10 and 90 (the default is 50).

A previously undocumented option that affects columns is `counts`. When this option is specified, more columns will appear on the right of the graph displaying the raw data; either the 2×2 table for binary data or the sample size, mean, and standard deviation in each group if the data are continuous. The groups may be labeled by using `group1(`*string*`)` and `group2(`*string*`)`, although the defaults *Treatment* and *Control* will often be acceptable for the analysis of randomized controlled trials (RCTs).

▷ **Example**

We now present an example command that uses these features, as well as the `second()` option. The resulting forest plot is displayed in figure 2:

```
. metan tcases tnoncases ccases cnoncases, rr fixedi second(random)
> lcols(trialnam authors startyr alloc latitude) counts astext(70)
> textsize(200) boxsca(80) xlabel(0.1,10) notable xsize(10) ysize(6)
```

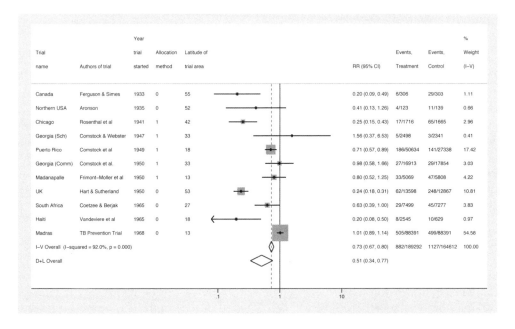

Figure 2. Forest plot displaying an inverse-variance weighted fixed-effects meta-analysis of the effect of BCG vaccine on incidence of tuberculosis. Columns of data are displayed in the plot.

Note the specification of x-axis labels and text and box sizes. The graph is also reshaped by using the standard Stata graph options `xsize()` and `ysize()`; see section 10.2 for more details. Box and text sizes are expressed as a percentage of standard size with the default as 100, such that 50 will halve the size and 200 will double it.

◁

6   by() processing

A major addition to `metan` is the ability to perform stratified or subgroup analyses. These may be used to investigate the possibility that treatment effects vary between subgroups; however, formal comparisons between subgroups are best performed by using meta-regression; see Harbord and Higgins (2008) or Higgins and Thompson (2004). We may also want to display results for different groups of studies in the same plot, even though it is inappropriate to meta-analyze across these groups.

6.1 Syntax and options for by()

nooverall specifies that the overall estimate not be displayed, for example, when it is inappropriate to meta-analyze across groups.

sgweight requests that weights be displayed such that they sum to 100% within each subgroup. This option is invoked automatically with nooverall.

nosubgroup specifies that studies be arranged by the subgroup specified, but estimates for each subgroup not be displayed.

nosecsub specifies that subestimates using the method defined by second() not be displayed.

summaryonly specifies that individual study estimates not be displayed, for example, to produce a summary of different groups in a compact graph.

▷ **Example**

Fine (1995) suggested that there is a relationship between the effect of BCG and the latitude of the area in which the trial was conducted. Here we may want to use meta-regression to further investigate this tendency (see Harbord and Higgins 2008). To illustrate the by() option, we will classify the studies into three groups defined by latitude. We define these groups as tropical (≤ 23.5 degrees), midlatitude (between 23.5 and 40 degrees), and northern (≥ 40 degrees).

```
. gen lat_cat = ""
(11 missing values generated)
. replace lat_cat = "Tropical, < 23.5 latitude" if latitude <= 23.5
lat_cat was str1 now str27
(4 real changes made)
. replace lat_cat = "23.5-40 latitude" if latitude > 23.5 & latitude < 40
(3 real changes made)
. replace lat_cat = "Northern, > 40 latitude" if latitude >= 40 & latitude < .
(4 real changes made)
. assert lat_cat != ""
. label var lat_cat "Latitude region"
```

(Continued on next page)

```
. metan tcases tnoncases ccases cnoncases, rr fixedi second(random) nosecsub
> lcols(trialnam startyr latitude) astext(60) by(lat_cat) xlabel(0.1,10)
> xsize(10) ysize(8)
```

Study	RR	[95% Conf. Interval]		% Weight
Northern, > 40 lat				
Canada	0.205	0.086	0.486	1.11
Northern USA	0.411	0.134	1.257	0.66
Chicago	0.254	0.149	0.431	2.96
UK	0.237	0.179	0.312	10.81
Sub-total				
I-V pooled RR	0.243	0.193	0.306	15.54
23.5-40 latitude				
Georgia (Sch)	1.562	0.374	6.528	0.41
Georgia (Comm)	0.983	0.582	1.659	3.03
South Africa	0.625	0.393	0.996	3.83
Sub-total				
I-V pooled RR	0.795	0.567	1.114	7.27
Tropical, < 23.5 l				
Puerto Rico	0.712	0.573	0.886	17.42
Madanapalle	0.804	0.516	1.254	4.22
Haiti	0.198	0.078	0.499	0.97
Madras	1.012	0.895	1.145	54.58
Sub-total				
I-V pooled RR	0.904	0.815	1.003	77.19
Overall				
I-V pooled RR	0.730	0.667	0.800	100.00
D+L pooled RR	0.508	0.336	0.769	

Test(s) of heterogeneity:

	Heterogeneity statistic	degrees of freedom	P	I-squared**
Northern, > 40 lat	1.06	3	0.787	0.0%
23.5-40 latitude	2.51	2	0.285	20.2%
Tropical, < 23.5 l	18.42	3	0.000	83.7%
Overall	125.63	10	0.000	92.0%

Overall Test for heterogeneity between sub-groups:

103.64	2	0.000

** I-squared: the variation in RR attributable to heterogeneity)

Considerable heterogeneity observed (up to 83.7%) in one or more sub-groups,
Test for heterogeneity between sub-groups likely to be invalid

Significance test(s) of RR=1

Northern, > 40 lat	z= 12.00	p = 0.000
23.5-40 latitude	z= 1.33	p = 0.183
Tropical, < 23.5 l	z= 1.90	p = 0.058
Overall	z= 6.75	p = 0.000

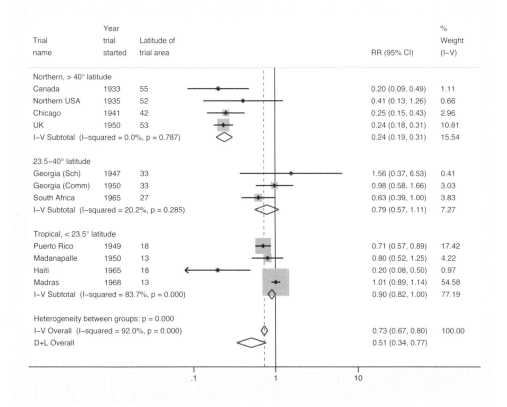

Figure 3. Forest plot displaying an inverse-variance weighted fixed-effects meta-analysis of the effect of BCG vaccine on incidence of tuberculosis. Results are stratified by latitude region, and the overall random-effects estimate is also displayed.

The output table is now stratified by latitude group, and pooled estimates for each group are displayed. Tests of heterogeneity and the null hypothesis are displayed for each group and overall. With the inverse-variance method, a test of heterogeneity between groups is also displayed; note the warning in the output that the test may be invalid because of within-subgroup heterogeneity. Output is similar in the forest plot, displayed in figure 3. Examining each subgroup in turn, it appears that much of the heterogeneity is accounted for by latitude: for two of the groups there is little or no evidence of heterogeneity. The only group to show a strong treatment effect is the ≥40 degree group.

◁

The test for between-group heterogeneity is an issue of current debate, as it is strictly valid only when using the fixed-effects inverse-variance method, and p-values will be too small if there is heterogeneity within any of the subgroups. Therefore, the test is performed only with the inverse-variance method (`fixedi`), and warnings will appear if there is evidence of within-group heterogeneity. Despite these caveats, this method is better than other, seriously flawed, methods such as testing the significance of a treatment effect in each group rather than testing for differences between the groups. As explained at the start of this section, meta-regression is the best way to examine and test for between-group differences.

7   User-defined analyses

7.1   Study weights

The `wgt(`*wgtvar*`)` option allows the studies to be combined by using specific weights that are defined by the variable *wgtvar*. The user must ensure that the weights chosen are meaningful. Typical uses are when analyzing precalculated effect estimates that require weights that are not based on standard error or to assess the robustness of conclusions by assigning alternative weights.

7.2   Pooled estimates

Pooled estimates may be derived by using another package and presented in a forest plot by using the `first()` option to supply these to the `metan` command. Here `wgt(`*wgtvar*`)` is used merely to specify box sizes in the forest plot, no heterogeneity statistics are produced, and no values are returned. When using this feature, stratified analyses are not allowed.

An alternative method is to provide the user-supplied meta-analytic estimate by using the `second()` option. Data are analyzed by using standard methods, and the resulting pooled estimate is displayed together with the user-defined estimate (which need not be derived by using `metan`), allowing a comparison. When using this feature, the option `nosecsub` is invoked, as stratification using the user-defined method is not possible.

When these options are specified, the user must supply the pooled estimate with its standard error or CI and a method label. The user may also supply text to be displayed at the bottom of the forest plot, in the position normally given to heterogeneity statistics, using `firststats(`*string*`)` and `secondstats(`*string*`)`.

▷ **Example**

The BCG data were analyzed by using a fully Bayesian random-effects model with WinBUGS software (Lunn et al. 2000). This analysis used the methods described by Warn, Thompson, and Spiegelhalter (2002) to deal with RRs. The chosen model incorporated a noninformative prior (mean 0, precision 0.001). The resulting RR of 0.518 (95% CI: 0.300, 0.824) is similar to that derived from a DerSimonian and Laird random-effects analysis. However, the CI from the Bayesian analysis is wider, because it allows for the uncertainty in estimating the between-study variance. The following syntax supplies the summary estimates in `second()` and compares this result with the random-effects analysis. The resulting forest plot is displayed in figure 4.

```
. metan logRR selogRR, random second(-.6587 -1.205 -.1937 Bayes)
> secondstats(Noninformative prior: d~dnorm(0.0, 0.001)) eform
> notable astext(60) textsize(130) lcols(trialnam startyr latitude)
> xlabel(0.1,10)
```

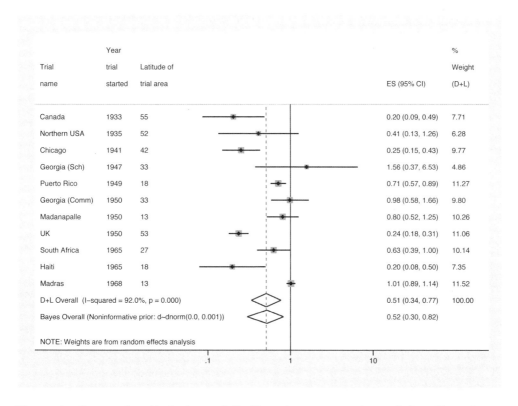

Figure 4.   Forest plot displaying a fully Bayesian meta-analysis of the effect of BCG vaccine on incidence of tuberculosis. A noninformative prior has been specified, resulting in a pooled-effect estimate similar to the random-effects analysis.

◁

8   New analysis options

Here we discuss previously undocumented options added to `metan` since its original publication.

8.1 Dealing with zero cells

The cc(#) option allows the user to choose what value (if any) is to be added to the cells of the 2×2 table for a study in which one or more of the cell counts equals zero. Here the default is to add 0.5 to all cells of the 2×2 table for the study (except for the Peto method, which does not require a correction). This approach has been criticized, and other approaches (including making no correction) may be preferable (see Sweeting, Sutton, and Lambert [2004] for a discussion). The number declared in cc(#) must be between zero and one and will be added to each cell. When no events are recorded and RRs or odds ratios are to be combined the study is omitted, although for risk differences the effect is still calculable and the study is included. If no adjustment is made in the presence of zero cells, odds ratios and their standard errors cannot be calculated. Risk ratios and their standard errors cannot be calculated when the number of events in either the treatment or control group is zero.

8.2 Noninteger sample size

The nointeger option allows the number of observations in each arm (cell counts for binary data or the number of observations for continuous data) to be noninteger. By default, the sample size is assumed to be a whole number for both binary and continuous data. However, it may make sense for this not to be so, for example, to use a more flexible continuity correction with a different number added to each cell or when the meta-analysis incorporates cluster randomized trials and the effective-sample size is less than the total number of observations.

8.3 Breslow and Day test for heterogeneity

The breslow option can be used to perform the Breslow–Day test for heterogeneity of the odds ratio (Breslow and Day 1993). A review article by Reis, Hirji, and Afifi (1999) compared several different tests of heterogeneity and found this test to perform well in comparison to other asymptotic tests.

9 New output

9.1 The I^2 statistic

metan now displays the I^2 statistic as well as Cochran's Q to quantify heterogeneity, based on the work by Higgins and Thompson (2004) and Higgins et al. (2003). Briefly, I^2 is the percentage of variation attributable to heterogeneity and is easily interpretable. Cochran's Q can suffer from low power when the number of studies is low or excessive power when the number of studies is large. I^2 is calculated from the results of the meta-analysis by

$$I^2 = 100\% \times \frac{(Q - \mathrm{df})}{Q}$$

where Q is Cochran's heterogeneity statistic and df is the degrees of freedom. Negative values of I^2 are set to zero so that I^2 lies between 0% and 100%. A value of 0% indicates no observed heterogeneity, and larger values show increasing heterogeneity. Although there can be no absolute rule for when heterogeneity becomes important, Higgins et al. (2003) tentatively suggest adjectives of low for I^2 values between 25%–50%, moderate for 50%–75%, and high for $\geq 75\%$.

9.2   Prediction interval for the random-effects distribution

The presentation of summary random-effects estimates may sometimes be misleading, as the CI refers to the average true treatment effect, but this is assumed under the random-effects model to vary between studies. A CI derived from a larger number of studies exhibiting a high degree of heterogeneity could be of similar width to a CI derived from a smaller number of more homogeneous studies, but in the first situation, we will be much less sure of the range within which the treatment effect in a new study will lie (Higgins and Thompson 2001). The prediction interval for the treatment effect in a new trial may be approximated by using the formula

$$\mathrm{mean} \pm t_{\mathrm{df}} \times \sqrt{(\mathrm{se}^2 + \tau^2)}$$

where t is the appropriate centile point (e.g., 95%) of the t distribution with $k-2$ degrees of freedom, se^2 is the squared standard error, and τ^2 the between-study variance. This incorporates uncertainty in the location and spread of the random-effects distribution. The approximate prediction interval can be displayed in the forest plot, with lines extending from the summary diamond, by using the option `rfdist`. With ≤ 2 studies, the distribution is inestimable and effectively infinite; thus the interval is displayed with dotted lines. When heterogeneity is estimated to be zero, the prediction interval is still slightly wider than the summary diamond as the t statistic is always greater than the corresponding normal deviate. The coverage (e.g., 90%, 95%, or 99%) for the interval may be set by using the command `rflevel(#)`.

▷ **Example**

Here we display the prediction intervals corresponding to the stratified analyses derived in section 6.1. The resulting forest plot is displayed in figure 5.

```
. metan tcases tnoncases ccases cnoncases, rr random rfdist
> lcols(trialnam startyr latitude) astext(60) by(lat_cat) xlabel(0.1,10)
> xsize(10) ysize(8) notable
```

(Continued on next page)

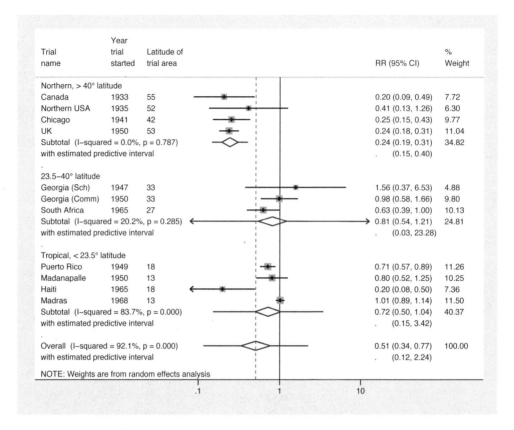

Figure 5. Forest plot displaying a random-effects meta-analysis of the effect of BCG vaccine on incidence of tuberculosis. Results are stratified by latitude region and the prediction interval for a future trial is displayed for each and overall.

◁

9.3 Vaccine efficacy

Results from the analysis of 2×2 data from vaccine trials may be reexpressed as the *vaccine efficacy* (also known as the *relative-risk reduction*); defined as the proportion of cases that would have been prevented in the placebo group had they received the vaccination (Kirkwood and Sterne 2001). The formula is

$$\text{Vaccine efficacy (VE)} = 100\% \times \left(1 - \frac{\text{risk of disease in vaccinated}}{\text{risk of disease in unvaccinated}}\right)$$

$$= 100\% \times (1 - \text{RR})$$

In `metan`, data are entered in the same way as any other analysis of 2×2 data and the option `efficacy` added. Results are displayed as odds ratios or RRs in the table and forest plot, but another column is added to the plot showing the results reexpressed as vaccine efficacy.

▷ **Example**

The BCG data are reanalyzed here, with results also displayed in terms of vaccine efficacy. The resulting forest plot is displayed in figure 6.

```
. metan tcases tnoncases ccases cnoncases, rr random efficacy
> lcols(trialnam startyr) textsize(150) notable xlabel(0.1, 10)
```

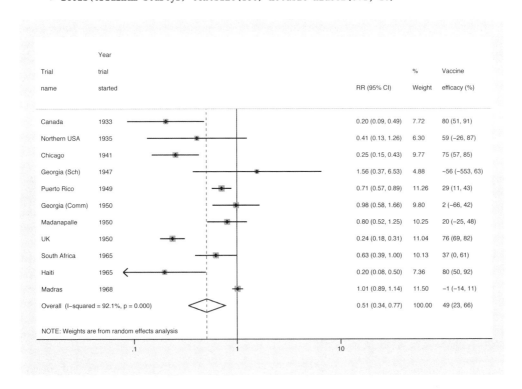

Figure 6. Forest plot displaying a random-effects meta-analysis of the effect of BCG vaccine on incidence of tuberculosis. Results are also displayed in terms of vaccine efficacy; estimates with a RR of greater than 1 produce a negative vaccine efficacy.

10   More graph options

10.1   metan graph options

Previous users of `metan` may find that they do not like the new box style and prefer a solid black box without the point estimate marker. The option `classic` changes back to this style. There are also options available to change the boxes, diamonds, and other lines. This is achieved by using options that change the standard graph commands that `metan` uses. For instance, the vertical line representing the overall effect may be changed using `olineopt()`, which can take standard Stata *line_options* such as `lwidth()`, `lcolor()`, and `lpattern()`. Boxes are weighted markers and not much can be changed, although shape and color may be modified by using *marker_options* in the `boxopt()` option, such as `msymbol()` and `mcolor()`, or we can dispense with the boxes entirely by using the option `nobox`. The point estimate markers have more flexibility and may also be modified by using *marker_options* in the `pointopt()` option; for instance, labels may by attached to them by using `mlabel()`. The CIs and diamonds may be changed by using *line_options* in the options `ciopt()` and `diamopt()`. For more details, see the `metan` help file and the Stata *Graphics Reference Manual* ([G] **graph**).

▷ **Example**

Here many aspects of the graph are changed and a raw data variable is defined (as in `counts`) and attached to the point estimates in the graph. The resulting graph is not shown here, but a similar application is shown in section 10.3.

```
. gen counts = string(tcases) + "/" + string(tcases+tnoncases) + "," +
> string(ccases) + "/" + string(ccases+cnoncases)
. metan tcases tnoncases ccases cnoncases, rr fixedi second(random) nosecsub
> notable olineopt(lwidth(thick) lcolor(navy) lpattern(dot))
> boxopt(msymbol(triangle) mcolor(dkgreen))
> pointopt(mlabel(counts) mlabsize(tiny) mlabposition(5))
```

◁

10.2   Overall graph options

Any graph options that come under the *overall*, *note*, and *caption* sections of Stata's `graph twoway` command may be added to a `metan` command, and the x axis (and y axis if required) may have a title added. The options `aspect()` or `xsize()` and `ysize()` may be used to specify different aspect ratios (e.g., portrait). The default aspect ratio of a Stata graph is around 0.7 (height/width), and `metan` tries to stick to this shape; although graphs that are more naturally displayed as long or wide will be reshaped to some degree. Use of the above options will control this more precisely.

Finally, the use of schemes is also supported. As colors of boxes and so on are defined within `metan`, these will not always give the desired result but may produce some interesting effects. Try, for example, using the scheme `economist`. More on schemes can be found in [G] **schemes intro**.

10.3   Notes on graph building

It can be useful to declare local or global macros that contain portions of code that are frequently used. For example, if the forest plot always has triangular "boxes" in forest green, contains the same columns of data, and so on, global macros may be declared for these bits of code. These can then be reused for a series of meta-analyses to specify the look and contents of the graphs. These could also be declared in an ado-file so that they are ready to use in every Stata session. This idea is similar to using Stata graph schemes.

▷ **Example**

Macros are defined to control various aspects of the graph and then used in the `metan` command. The resulting forest plot is displayed in figure 7.

```
. global metamethod rr fixedi second(random) nosecsub
. global metacolumns lcols(trialnam startyr latitude) astext(60)
. global metastyle boxopt(mcolor(forest_green) msymbol(triangle))
> pointopt(msymbol(smtriangle) mcolor(gold) msize(tiny)
> mlabel(counts) mlabsize(tiny) mlabposition(2) mlabcolor(brown))
> diamopt(lcolor(black) lwidth(medthick)) graphregion(fcolor(gs10)) boxsca(80)
. global metaopts favours(decreases TB # increases TB)
> xlabel(0.1, 0.2, 0.5, 2, 5, 10) notable
. metan tcases tnoncases ccases cnoncases,
> $metamethod $metacolumns $metastyle $metaopts by(lat_cat) xsize(10) ysize(8)
```

(Continued on next page)

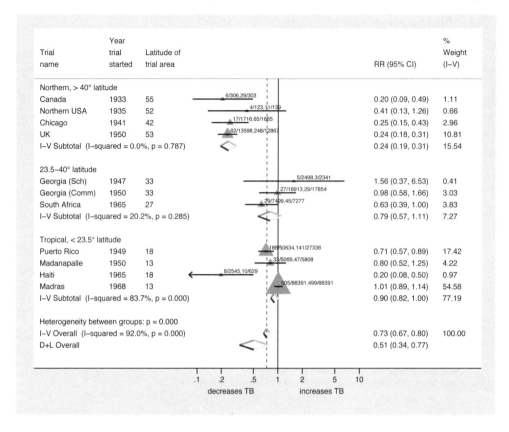

Figure 7. Forest plot displaying an inverse-variance weighted fixed-effects meta-analysis of the effect of BCG vaccine on incidence of tuberculosis. Results are stratified by latitude region, and the overall random-effects estimate is also displayed. Various options have been used to change the display of the graph.

◁

11 Variables and results produced by metan

11.1 Variables generated

When odds ratios (OR) or RRs are combined from 2×2 data and the log option is not used, the SE log OR or log RR is saved in a variable named _selogES, to make clear that it is the SE log OR or RR and not on the same scale. If the log option is used, the standard error is named _seES, as it is on the same scale as the estimate itself. In both cases, the estimate is called _ES.

It is possible to calculate the standard error of ORs and RRs by the delta method; this is what Stata does, for example, with the results reported by the logistic command.

However, the distribution of ratios is in general highly skewed, and for this reason, `metan` does not attempt to record the standard error of either the OR or RR.

Absolute measures (risk differences or mean differences) are symmetric and may be assumed to be normally distributed via the central limit theorem. Here `metan` stores these quantities in _ES and their standard errors in _seES. The derived variables incorporate the correction for zero cells (see section 8.1).

_ES	Effect size (ES)
_seES	Standard error of ES
_selogES	Standard error of log ES
_LCI	Lower confidence limit for ES
_UCI	Upper confidence limit for ES
_WT	Study percentage weight
_SS	Study sample size

11.2 Saved results (macros)

As with many Stata commands, macros are left behind containing the results of the analysis. If two methods are specified by using the option `second()`, some of these are repeated; for example, `r(ES)` and `r(ES_2)` give the pooled-effects estimates for each method. Subgroup statistics when using the `by()` option are not saved; if these are required for storage, it is recommended that a program be written that analyzes subgroups separately (perhaps using the `nograph` and `notable` options).

(Continued on next page)

Name	Second	Description
r(ES)	r(ES_2)	pooled-effect size (if the log option is specified with or or rr, this is the pooled log OR or log RR)
r(seES)	r(seES_2)	standard error of pooled-effect size with symmetrical CI, i.e., mean differences, risk difference, log OR, and log RR using log option
r(selogES)	r(selogES_2)	standard error of log OR or log RR when ORs or RRs are combined without the log option
r(ci_low)	r(ci_low_2)	lower CI of pooled-effect size
r(ci_upp)	r(ci_upp_2)	upper CI of pooled-effect size
r(z)		z-value of effect size
r(p_z)		p-value for significance of effect size
r(het)		chi-squared test for heterogeneity
r(df)		degrees of freedom (number of informative studies minus 1)
r(p_het)		p-value for significance of test for heterogeneity
r(i_sq)		the I^2 statistic
r(tau2)		estimated between-study variance (random-effects analyses only)
r(chi2)		chi-squared test for significance of odds ratio (fixed-effects OR only)
r(p_chi2)		p-value for the above test
r(rger)		overall event rate, group 1 (if binary data are combined)
r(cger)		overall event rate, group 2 (see above)
r(measure)		effect measure (e.g., RR, SMD)
r(method_1)	r(method_2)	analysis method (e.g., M-H, D+L)

12 References

Bradburn, M. J., J. J. Deeks, and D. G. Altman. 1998. sbe24: metan—an alternative meta-analysis command. *Stata Technical Bulletin* 44: 4–15. Reprinted in *Stata Technical Bulletin Reprints*, vol. 8, pp. 86–100. College Station, TX: Stata Press. (Updated article is reprinted in this collection on pp. 3–28.)

Breslow, N. E., and N. E. Day. 1993. *Statistical Methods in Cancer Research: Volume I—The Analysis of Case–Control Studies.* Lyon, UK: International Agency for Research on Cancer.

Colditz, G. A., T. F. Brewer, C. S. Berkey, M. E. Wilson, E. Burdick, H. V. Fineberg, and F. Mosteller. 1994. Efficacy of BCG vaccine in the prevention of tuberculosis.

Meta-analysis of the published literature. *Journal of the American Medical Association* 271: 698–702.

Deeks, J. J., D. G. Altman, and M. J. Bradburn. 2001. Statistical methods for examining heterogeneity and combining results from several studies in meta-analysis. In *Systematic Reviews in Health Care: Meta-analysis in Context*, 2nd edition, ed. M. Egger, G. Davey Smith, and D. G. Altman, 285–321.

Fine, P. E. M. 1995. Variation in protection by BCG: Implications of and for heterologous immunity. *Lancet* 346: 1339–1345.

Harbord, R. M., and J. P. T. Higgins. 2008. Meta-regression in Stata. *Stata Journal* 8: 493–519. (Reprinted in this collection on pp. 70–96.)

Higgins, J. P. T., and S. G. Thompson. 2001. Presenting random effects meta-analyses: Where are we going wrong? In *9th International Cochrane Colloquium*. Lyon, France.

———. 2004. Controlling the risk of spurious findings from meta-regression. *Statistics in Medicine* 23: 1663–1682.

Higgins, J. P. T., S. G. Thompson, J. J. Deeks, and D. G. Altman. 2003. Measuring inconsistency in meta-analyses. *British Medical Journal* 327: 557–560.

Kirkwood, B. R., and J. A. C. Sterne. 2001. *Essentials of Medical Statistics*. 2nd ed. Oxford: Blackwell Science.

Lunn, D. J., A. Thomas, N. Best, and D. Spiegelhalter. 2000. WinBUGS – A Bayesian modelling framework: Concepts, structure and extensibility. *Statistics and Computing* 10: 325–337.

Poole, C., and S. Greenland. 1999. Random-effects meta-analyses are not always conservative. *American Journal of Epidemiology* 150: 469–475.

Reis, I. M., K. F. Hirji, and A. A. Afifi. 1999. Exact and asymptotic tests for homogeneity in several 2×2 tables. *Statistics in Medicine* 18: 893–906.

Sharp, S., and J. A. C. Sterne. 1997. sbe16: Meta-analysis. *Stata Technical Bulletin* 38: 9–14. Reprinted in *Stata Technical Bulletin Reprints*, vol. 7, pp. 100–106. College Station, TX: Stata Press.[1]

———. 1998. sbe16.1: New syntax and output for the meta-analysis command. *Stata Technical Bulletin* 42: 6–8. Reprinted in *Stata Technical Bulletin Reprints*, vol. 7, pp. 106–108. College Station, TX: Stata Press.[1]

Sterne, J. A. C., D. Gavaghan, and M. Egger. 2000. Publication and related bias in meta-analysis: Power of statistical tests and prevalence in the literature. *Journal of Clinical Epidemiology* 53: 1119–1129.

---

1. The original command to perform meta-analysis was `meta`, documented in the sbe16 articles; `meta` is now `metan`. `metan` is described in an updated article, sbe24, on pages 3–28 of this collection.—Ed.

Sweeting, M. J., A. J. Sutton, and P. C. Lambert. 2004. What to add to nothing? Use and avoidance of continuity corrections in meta-analysis of sparse data. *Statistics in Medicine* 23: 1351–1375.

Warn, D. E., S. G. Thompson, and D. J. Spiegelhalter. 2002. Bayesian random effects meta-analysis of trials with binary outcomes: Methods for absolute risk difference and relative risk scales. *Statistics in Medicine* 21: 1601–1623.

Cumulative meta-analysis[1]

Jonathan A. C. Sterne
Department of Social Medicine
University of Bristol
Bristol, UK

Meta-analysis is used to combine the results of several studies, and the Stata command metan (Bradburn, Deeks, and Altman 1998; Harris et al. 2008)[2] can be used to perform meta-analyses and graph the results. In cumulative meta-analysis (Lau et al. 1992), the pooled estimate of the treatment effect is updated each time the results of a new study are published. This makes it possible to track the accumulation of evidence on the effect of a particular treatment.

The command metacum performs cumulative meta-analysis (using fixed- or random-effects models) and, optionally, graphs the results.

1 Syntax

metacum *varlist* [*if*] [*in*] [,

[*binary_data_options* | *continuous_data_options* | *precalculated_effect_estimates_options*]

measure_and_model_option output_options forest_plot_options]

binary_data_options

> or rr rd fixed random fixedi randomi peto <u>noint</u>eger cc(#)

continuous_data_options

> cohen hedges glass nostandard fixed random <u>noint</u>eger

precalculated_effect_estimates_options

> fixed random

measure_and_model_option

> wgt(*wgtvar*)

---

1. The updated syntax is by Ross Harris, Centre for Infections, Health Protection Agency, London.—
Ed.
2. The original command to be installed was meta; see Sharp and Sterne (1997, 1998).—Ed.

output_options

```
by(byvar) log eform ilevel(#)
sortby(varlist)
label([namevar = namevar], [yearvar = yearvar]) notable nograph
```

forest_plot_options

```
xlabel(#, ...) xtick(#, ...) textsize(#) nowt nostats counts
group1(string) group2(string) effect(string) force lcols(varlist)
rcols(varlist) astext(#) double summaryonly null(#) nulloff
favours(string # string)
pointopt(marker_options | marker_label_options) ciopt(line_options)
olineopt(line_options) classic nowarning graph_options
```

2 Options

2.1 binary_data_options

or pools ORs.

rr pools RRs; this is the default.

rd pools risk differences.

fixed specifies a fixed-effects model using the Mantel–Haenszel method; this is the default.

random specifies a random-effects model using the DerSimonian and Laird method, with the estimate of heterogeneity being taken.

fixedi specifies a fixed-effects model using the inverse-variance method.

randomi specifies a random-effects model using the DerSimonian and Laird method, with the estimate of heterogeneity being taken from the inverse-variance fixed-effects model.

peto specifies that the Peto method is used to pool ORs.

nointeger allows the cell counts to be nonintegers. This option may be useful when a variable continuity correction is sought for studies containing zero cells but also may be used in other circumstances, such as where a cluster-randomized trial is to be incorporated and the "effective sample size" is less than the total number of observations.

cc(#) defines a fixed-continuity correction to add where a study contains a zero cell. By default, metan8 adds 0.5 to each cell of a trial where a zero is encountered when using inverse-variance, DerSimonian and Laird, or Mantel–Haenszel weighting to enable finite variance estimators to be derived. However, the cc() option allows the use of other constants (including none). See also the **nointeger** option.

2.2 continuous_data_options

cohen pools standardized mean differences by the Cohen method; this is the default.

hedges pools standardized mean differences by the Hedges method.

glass pools standardized mean differences by the Glass method.

nostandard pools unstandardized mean differences.

fixed specifies a fixed-effects model using the Mantel–Haenszel method; this is the default.

random specifies a random-effects model using the DerSimonian and Laird method, with the estimate of heterogeneity being taken.

nointeger denotes that the number of observations in each arm does not need to be an integer. By default, the first and fourth variables specified (containing N_intervention and N_control, respectively) may occasionally be noninteger (see **nointeger** in section 2.1).

2.3 precalculated_effect_estimates_options

fixed specifies a fixed-effects model using the Mantel–Haenszel method; this is the default.

random specifies a random-effects model using the DerSimonian and Laird method, with the estimate of heterogeneity being taken.

2.4 measure_and_model_option

wgt(*wgtvar*) specifies alternative weighting for any data type. The effect size is to be computed by assigning a weight of *wgtvar* to the studies. When RRs or ORs are declared, their logarithms are weighted. This option should be used only if you are satisfied that the weights are meaningful.

2.5 output_options

by(*byvar*) specifies that the meta-analysis is to be stratified according to the variable declared.

log reports the results on the log scale (valid only for ORs and RRs analyses from raw data counts).

eform exponentiates all effect sizes and confidence intervals (valid only when the input variables are log ORs or log hazard-ratios with standard error or confidence intervals).

ilevel(*#*) specifies the coverage (e.g., 90%, 95%, 99%) for the individual trial confidence intervals; the default is $S_level. **ilevel()** and **olevel()** need not be the same. See [U] **20.7 Specifying the width of confidence intervals**.

sortby(*varlist*) sorts by variable(s) in *varlist*.

label([namevar=*namevar*], [yearvar=*yearvar*]) labels the data by its name, year, or both. Either or both variable lists may be left blank. For the table display, the overall length of the label is restricted to 20 characters. If the lcols() option is also specified, it will override the label() option.

notable prevents the display of a results table.

nograph prevents the display of a graph.

2.6   forest_plot_options

xlabel(#,...) defines *x*-axis labels. This option has been modified so that any number of points may be defined. Also, checks are no longer made as to whether these points are sensible, so the user may define anything if the force option is used. Points must be comma separated.

xtick(#,...) adds tick marks to the *x* axis. Points must be comma separated.

textsize(#) specifies the font size for the text display on the graph. This option has been modified so that the default is textsize(100) (as in 100%) and the percentage may be increased or decreased (e.g., 80 or 120 for 20% smaller or larger, respectively).

nowt prevents the display of study weight on the graph.

nostats prevents the display of study statistics on the graph.

counts displays data counts (n/N) for each group when using binary data or the sample size, mean, and standard deviation for each group if mean differences are used (the latter is a new feature).

group1(*string*) and group2(*string*) may be used with the counts option, and the text should contain the names of the two groups.

effect(*string*) allows the graph to name the summary statistic used when the effect size and its standard error are declared.

force forces the x-axis scale to be in the range specified by xlabel().

lcols(*varlist*) and rcols(*varlist*) define columns of additional data to the left or right of the graph. The first two columns on the right are automatically set to effect size and weight, unless suppressed by using the options nostats and nowt. If counts is used, this will be set as the third column. textsize() can be used to fine-tune the size of the text to achieve a satisfactory appearance. The columns are labeled with the variable label or the variable name if this is not defined. The first variable specified in lcols() is assumed to be the study identifier and this is used in the table output.

astext(#) specifies the percentage of the graph to be taken up by text. The default is 50%, and the percentage must be in the range 10–90.

double allows variables specified in lcols() and rcols() to run over two lines in the plot. This option may be of use if long strings are used.

summaryonly shows only summary estimates in the graph. This option may be of use for multiple subgroup analyses.

null(#) displays the null line at a user-defined value rather than at 0 or 1.

nulloff removes the null hypothesis line from the graph.

favours(*string # string*) applies a label saying something about the treatment effect to either side of graph (strings are separated by the # symbol). This option replaces the feature available in b1title in the previous version of metan.

pointopt(*marker_options | marker_label_options*), ciopt(*line_options*), and olineopt(*line_options*) specify options for the graph routines within the program, allowing the user to alter the appearance of the graph. Any options associated with a particular graph command may be used, except some that would cause incorrect graph appearance. For example, diamonds are plotted using the twoway pcspike command, so options for line styles are available (see [G] *line_options*); however, altering the $x - -y$ orientation with the option horizontal or vertical is not allowed. So, ciopt(lcolor(green) lwidth(thick)) feeds into a command such as pcspike(y1 x1 y2 x2, lcolor(green) lwidth(thick))

> pointopt(*marker_options | marker_label_options*), controls the point estimate by using marker options. See [G] *marker_options* and [G] *marker_label_options*.
>
> ciopt(*line_options*) controls the confidence intervals for studies by using options for twoway pcspike (not horizontal/vertical). See [G] *line_options*.
>
> olineopt(*line_options*) controls the overall effect line with options for another line (not position). See [G] *line_options*.

classic specifies that solid black boxes without point estimate markers are used, as in the previous version of metan.

nowarning switches off the default display of a note warning that studies are weighted from random-effects analyses.

graph_options are any of the options documented in [G] *twoway_options*. These allow the addition of titles, subtitles, captions, etc.; control of margins, plot regions, graph size, aspect ratio; and the use of schemes. Because titles may be added with *graph_options*, previous options such as b2title are no longer necessary.

3   Background

The command metacum provides an alternative means of presenting the results of a meta-analysis, where instead of the individual study effects and combined estimate, the cumulative evidence up to and including each trial can be printed and/or graphed. The technique was suggested by Lau et al. (1992).

4   Example

The first trial of streptokinase treatment following myocardial infarction was reported in 1959. A further 21 trials were conducted between that time and 1986, when the ISIS-2 multicenter trial (on over 17,000 patients in whom over 1800 deaths were reported) demonstrated conclusively that the treatment reduced the chances of subsequent death.

Lau et al. (1992) pointed out that a meta-analysis of trials performed up to 1977 provided strong evidence that the treatment worked. Despite this, it was another 15 years until the treatment became routinely used.

Dataset `strepto.dta` contains the results of 22 trials of streptokinase conducted between 1959 and 1986.

```
. describe
Contains data from streptok.dta
  obs:            22                          Streptokinase and CHD
  vars:            7                          8 Jun 2005 15:47
  size:          638 (99.9% of memory free)   (_dta has notes)

              storage   display    value
variable name   type    format     label      variable label

trial          byte     %8.0g                 trial number
trialnam       str14    %14s                  Trial name
year           int      %8.0g                 year published
pop1           int      %12.0g                Treated population
deaths1        int      %12.0g                Treated cases
pop0           int      %12.0g                Control population
deaths0        int      %12.0g                Control cases

Sorted by:  trial

. list trialnam year pop1 deaths1 pop0 deaths0, noobs clean
           trialnam   year   pop1   deaths1   pop0   deaths0
           Fletcher   1959     12         1     11         4
              Dewar   1963     21         4     21         7
       1st European   1969     83        20     84        15
        Heikinheimo   1971    219        22    207        17
            Italian   1971    164        19    157        18
       2nd European   1971    373        69    357        94
       2nd Frankfurt  1973    102        13    104        29
      1st Australian  1973    264        26    253        32
          NHLBI SMIT   1974     53         7     54         3
             Valere   1975     49        11     42         9
              Frank   1975     55         6     53         6
           UK Collab  1976    302        48    293        52
              Klein   1976     14         4      9         1
            Austrian  1977    352        37    376        65
           Lasierra   1977     13         1     11         3
           N German   1977    249        63    234        51
           Witchitz   1977     32         5     26         5
      2nd Australian  1977    112        25    118        31
       3rd European   1977    156        25    159        50
               ISAM   1986    859        54    882        63
            GISSI-1   1986   5860       628   5852       758
             ISIS-2   1988   8592       791   8595      1029
```

Before doing our meta-analysis, we calculate the log odds-ratio for each study, and its corresponding variance. We also create a string variable containing the trial name and year of publication:

```
. gen logor=log((deaths1/(pop1-deaths1))/((deaths0/(pop0-deaths0))))

. gen selogor=sqrt(1/deaths1+1/(pop1-deaths1)+1/deaths0+1/(pop0-deaths0))

. metan logor selogor, eform fixed label(namevar=trialnam, yearvar=year)
> xlabel(.1,.5,1,2,10) force effect("Odds ratio")
```

Study		ES	[95% Conf. Interval]		% Weight
Fletcher (1959)		0.159	0.015	1.732	0.07
Dewar (1963)		0.471	0.114	1.942	0.21
1st European (1969)		1.460	0.689	3.096	0.75
Heikinheimo (1971)		1.248	0.643	2.423	0.96
Italian (1971)		1.012	0.510	2.008	0.90
2nd European (1971)		0.635	0.447	0.903	3.42
2nd Frankfurt (1973)		0.378	0.183	0.778	0.81
1st Australian (1973		0.754	0.436	1.306	1.41
NHLBI SMIT (1974)		2.587	0.632	10.596	0.21
Valere (1975)		1.061	0.392	2.876	0.43
Frank (1975)		0.959	0.289	3.185	0.29
UK Collab (1976)		0.876	0.570	1.346	2.29
Klein (1976)		3.200	0.296	34.588	0.07
Austrian (1977)		0.562	0.365	0.867	2.26
Lasierra (1977)		0.222	0.019	2.533	0.07
N German (1977)		1.215	0.797	1.853	2.38
Witchitz (1977)		0.778	0.199	3.044	0.23
2nd Australian (1977		0.806	0.440	1.477	1.16
3rd European (1977)		0.416	0.242	0.716	1.44
ISAM (1986)		0.872	0.599	1.270	2.99
GISSI-1 (1986)		0.807	0.721	0.903	33.44
ISIS-2 (1988)		0.746	0.676	0.822	44.19
I-V pooled ES		0.774	0.725	0.826	100.00

```
  Heterogeneity chi-squared =  31.50 (d.f. = 21) p = 0.066
  I-squared (variation in ES attributable to heterogeneity) =  33.3%

  Test of ES=1 : z=  7.71 p = 0.000
```

(Continued on next page)

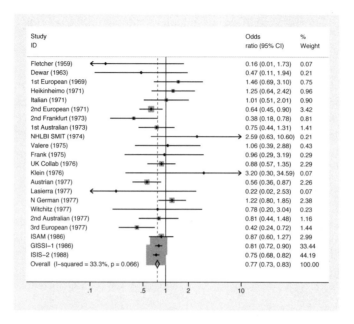

Figure 1. Streptokinase meta-analysis

It can be seen from the fixed-effects weights, and the graphical display, that the results are dominated by the two large trials reported in 1986. We now do a cumulative meta-analysis:

```
. metacum logor selogor, eform fixed label(namevar=trialnam, yearvar=year)
> xlabel(.1,.5,1,2) force effect("Odds ratio")
          Study        |   ES    [95% Conf. Interval]
----------------------+----------------------------------------------------
Fletcher (1959)       |  0.159     0.015      1.732
Dewar (1963)          |  0.355     0.105      1.200
1st European (1969)   |  0.989     0.522      1.875
Heikinheimo (1971)    |  1.106     0.698      1.753
Italian (1971)        |  1.076     0.734      1.577
2nd European (1971)   |  0.809     0.624      1.048
2nd Frankfurt (1973)  |  0.742     0.581      0.946
1st Australian (1973  |  0.744     0.595      0.929
NHLBI SMIT (1974)     |  0.767     0.615      0.955
Valere (1975)         |  0.778     0.628      0.965
Frank (1975)          |  0.783     0.634      0.968
UK Collab (1976)      |  0.801     0.662      0.968
Klein (1976)          |  0.808     0.668      0.976
Austrian (1977)       |  0.762     0.641      0.906
Lasierra (1977)       |  0.757     0.637      0.900
N German (1977)       |  0.811     0.691      0.951
Witchitz (1977)       |  0.810     0.691      0.950
2nd Australian (1977  |  0.810     0.695      0.945
3rd European (1977)   |  0.771     0.665      0.894
ISAM (1986)           |  0.784     0.683      0.899
GISSI-1 (1986)        |  0.797     0.731      0.870
ISIS-2 (1988)         |  0.774     0.725      0.826
----------------------+----------------------------------------------------
```

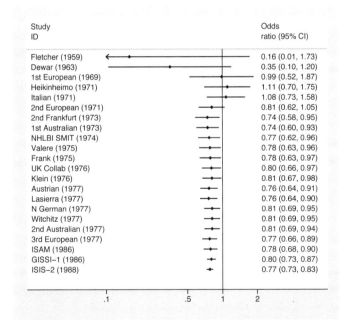

Figure 2. Streptokinase cumulative meta-analysis

By the end of 1977 there was clear evidence that streptokinase treatment prevented death following myocardial infarction. The point estimate of the pooled treatment effect was virtually identical in 1977 (odds ratio = 0.771) and after the results of the large trials in 1986 (odds ratio = 0.774).

5   Note

The command `metan` (Bradburn, Deeks, and Altman 1998; Harris et al. 2008)[3] should be installed before running `metacum`.

6   Acknowledgments

I thank Stephen Sharp for reviewing the command, Matthias Egger for providing the streptokinase data, and Thomas Steichen for providing the alternative forms of command syntax.

7   References

Bradburn, M. J., J. J. Deeks, and D. G. Altman. 1998. sbe24: metan—an alternative meta-analysis command. *Stata Technical Bulletin* 44: 4–15. Reprinted in *Stata Technical Bulletin Reprints*, vol. 8, pp. 86–100. College Station, TX: Stata Press. (Updated article is reprinted in this collection on pp. 3–28.)

Harris, R. J., M. J. Bradburn, J. J. Deeks, R. M. Harbord, D. G. Altman, and J. A. C. Sterne. 2008. metan: fixed- and random-effects meta-analysis. *Stata Journal* 8: 3–28. (Reprinted in this collection on pp. 29–54.)

Lau, J., E. M. Antman, J. Jimenez-Silva, B. Kupelnick, F. Mosteller, and T. C. Chalmers. 1992. Cumulative meta-analysis of therapeutic trials for myocardial infarction. *New England Journal of Medicine* 327: 248–254.

Sharp, S., and J. A. C. Sterne. 1997. sbe16: Meta-analysis. *Stata Technical Bulletin* 38: 9–14. Reprinted in *Stata Technical Bulletin Reprints*, vol. 7, pp. 100–106. College Station, TX: Stata Press.[4]

———. 1998. sbe16.1: New syntax and output for the meta-analysis command. *Stata Technical Bulletin* 42: 6–8. Reprinted in *Stata Technical Bulletin Reprints*, vol. 7, pp. 106–108. College Station, TX: Stata Press.[4]

---

3. The original command to be installed was `meta`; see Sharp and Sterne (1997, 1998).—Ed.

4. The original command to perform meta-analysis was `meta`, documented in the sbe16 articles; `meta` is now `metan`. `metan` is described in an updated article, sbe24, on pages 3–28 of this collection.—Ed.

The Stata Technical Bulletin (1999)
STB-49, pp. 15–17   65

Meta-analysis of p-values

Aurelio Tobias
Senior Medical Statistician
National School of Public Health
Instituto de Saluid Carlos III
Madrid, Spain
atobias@isciii.es

Fisher's work on combining p-values (Fisher 1932) has been suggested as the origin of meta-analysis (Jones 1995). However, combination of p-values presents serious disadvantages, relative to combining estimates. For example, when p-values are testing different null hypotheses, they do not consider the direction of the association combining opposing effects, they cannot quantify the magnitude of the association, nor study heterogeneity between studies. Combination of p-values may be the only available option if nonparametric analyses of individual studies have been performed or if little information apart from the p-value is available about the result of a particular study (Jones 1995).

1 Fisher's method

This method (Fisher 1932) combines the probabilities of several hypotheses tests, testing the same null hypothesis

$$U = -2 \sum_{j=1}^{k} \ln(p_j)$$

where the p_j are the one-tailed p-values for each study, and k is the number of studies. Then U follows a χ^2 distribution with $2k$ degrees of freedom. This method is not suggested to combine a large number of studies because it tends to reject the null hypothesis routinely (Rosenthal 1984). It also tends to have problems combining studies that are statistically significant, but in opposite directions (Rosenthal 1980).

2 Edgington's methods

The first method (Edgington 1972a) is based on the sum of probabilities

$$p = \left(\sum_{j=1}^{K} p_j \right)^{k} \Big/ k!$$

The results obtained are similar to Fisher's method, but it is also restricted for a small number of studies. This method presents problems when the sum of probabilities is higher than one; in this situation the combined probability tends to be conservative (Rosenthal 1980).

An alternative method was also suggested by Edgington (1972b), to combine more than four studies, based on the contrast of the p-value average

$$\bar{p} = \sum_{j=1}^{k} p_j \Big/ k$$

in which case $U = (0.5 - \bar{p})\sqrt{12}$ follows a normal distribution.

3 Syntax

The command `metap` works on a dataset containing the p-values for each study. The syntax is as follows:

`metap` *pvar* $\big[\,if\,\big]$ $\big[\,in\,\big]$ $\big[\,$, $\underline{\text{m}}$ethod($\#$)$\big]$

4 Option

`method(#)` combines the p-values using three available methods:

`method(f)`, Fisher's method. This is the default.

`method(ea)`, Edgington's additive method based on the sum of probabilities (Edgington 1972a). This method is suggested to combine a small number of studies, producing similar results as Fisher's method.

`method(en)`, Edgington's normal curve method, based on the contrast of the p-value average (Edgington 1972b). This method is suggested to combine a large number of studies.

5 Example

We consider data from seven placebo-controlled studies on the effect of aspirin in preventing death after myocardial infarction. Fleiss (1993) published an overview of these data. Let us assume that each study included in the meta-analysis is testing the same null hypothesis $H_0\colon \theta \leq 0$ versus the alternative $H_1\colon \theta > 0$. If the estimate of the log odds-ratio and its standard error is available, then one-tailed p-values can easily be generated using the `normprob` function:

```
. generate pvar=normprob(-logrr/logse)
. list studyid logrr logse pvar, noobs
  studyid     logrr     logse       pvar
    MCR-1    0.3289    0.1972    .0476728
      CDP    0.3853    0.2029    .0287845
    MRC-2    0.2192    0.1432    .0629185
     GASP    0.2229    0.2545    .1905599
    PARIS    0.2261    0.1876    .1140584
     AMIS   -0.1249    0.0981    .8985248
   ISIS-2    0.1112    0.0388    .0020786
```

In this situation, all methods to combine *p*-values produce similar results:

```
. metap pvar
Meta-analysis of p_values
-----------------------------------------------------------------
  Method             |   chi2         p_value     studies
---------------------+-------------------------------------------
  Fisher             | 38.938235     .00037283    7
-----------------------------------------------------------------

. metap pvar, e(a)
Meta-analysis of p_values
-----------------------------------------------------------------
  Method             |    .           p_value     studies
---------------------+-------------------------------------------
  Edgington, additive|    .          .00157658    7
-----------------------------------------------------------------

. metap pvar, e(n)
Meta-analysis of p_values
-----------------------------------------------------------------
  Method             |   Z            p_value     studies
---------------------+-------------------------------------------
  Edgington, Normal  | 2.8220842     .00238563    7
-----------------------------------------------------------------
```

These figures agree with the result obtained using the `meta` command introduced in Sharp and Sterne (1997, 1998) on a fixed effects ($z = 3.289$, $p = 0.001$) and random effects ($z = 2.093$, $p = 0.036$) models, respectively. However, the combination of *p*-values presents the serious limitations described previously.

6   Individual or frequency records

As for other meta-analysis commands, `metap` works on data contained in frequency records, one for each study or trial.

7   Saved results

`metap` saves the following results:

S_1	method used to combine the *p*-values
S_2	number of studies
S_3	statistic used to obtain the combined probability
S_4	values of the statistic described in S_3
S_5	combined probability
r(method)	the method used to combine the *p*-values
r(n)	the number of studies
r(stat)	the statistic used to combine the *p*-values
r(z)	the value of the statistic used
r(pvalue)	returns the combined *p*-value

8   References

Edgington, E. S. 1972a. An additive method for combining probability values from independent experiments. *Journal of Psychology* 80: 351–363.

———. 1972b. A normal curve method for combining probability values from independent experiments. *Journal of Psychology* 82: 85–89.

Fisher, R. A. 1932. *Statistical Methods for Research Workers.* 4th ed. London: Oliver & Boyd.

Fleiss, J. L. 1993. The statistical basis of meta-analysis. *Statistical Methods in Medical Research* 2: 121–145.

Jones, D. R. 1995. Meta-analysis: Weighing the evidence. *Statistics in Medicine* 14: 137–149.

Rosenthal, R., ed. 1980. *New Directions for Methodology of Social and Behavioral Science.* Volume V. San Francisco: Sage.

Rosenthal, R. 1984. Valid interpretation of quantitative research results. In *New Directions for Methodology of Social and Behavioral Science: Forms of Validity in Research,* ed. D. Brinberg and L. Kidder, 12. San Francisco: Jossey–Bass.

Sharp, S., and J. A. C. Sterne. 1997. sbe16: Meta-analysis. *Stata Technical Bulletin* 38: 9–14. Reprinted in *Stata Technical Bulletin Reprints*, vol. 7, pp. 100–106. College Station, TX: Stata Press.[1]

———. 1998. sbe16.1: New syntax and output for the meta-analysis command. *Stata Technical Bulletin* 42: 6–8. Reprinted in *Stata Technical Bulletin Reprints*, vol. 7, pp. 106–108. College Station, TX: Stata Press.[1]

---

1. The original command to perform meta-analysis was `meta`, documented in the sbe16 articles; `meta` is now `metan`. `metan` is described in an updated article, sbe24, on pages 3–28 of this collection.—Ed.

Part 2

Meta-regression: metareg

Interpretation of the results of meta-analyses is simplest when there is little between-study heterogeneity. It is then appropriate to report a fixed-effects analysis, which assumes that the studies estimate the same underlying effect. Random-effects meta-analyses allow for between-study heterogeneity, but the interpretation of such analyses is more subtle than is commonly realized because rather than estimating a single effect, they estimate a mean effect, while the true effect is assumed to vary between studies around this mean. It is desirable, rather than simply allowing for heterogeneity, to understand reasons for it. This can be done using meta-regression.

The user-written `metareg` command remains one of the few implementations of meta-regression and has been updated to take account of improvements in Stata estimation facilities and recent methodological developments. However, enthusiasm for meta-regression has been tempered over the last decade. These are observational analyses in which the unit of analysis is the study. There are often more potential explanations for heterogeneity than studies in the meta-analysis, and Higgins and Thompson (2004) have shown that, in this context, multiple comparisons can substantially inflate the rate of false-positive findings. The updated `metareg` command includes the permutation test proposed by these authors to deal with the problem of multiple comparisons.

Because the 2008 article by Harbord and Higgins describes a comprehensive update to the `metareg` command, this appears first; the original syntax continues to work. This syntax and the formulas implemented in the 1998 version of `metareg` are then described in the article by Sharp.

1   Reference

Higgins, J. P. T., and S. G. Thompson. 2004. Controlling the risk of spurious findings from meta-regression. *Statistics in Medicine* 23: 1663–1682.

The Stata Journal (2008)
8, Number 4, pp. 493–519

Meta-regression in Stata

Roger M. Harbord
Department of Social Medicine
University of Bristol
Bristol, UK
roger.harbord@bristol.ac.uk

Julian P. T. Higgins
MRC Biostatistics Unit
Cambridge, UK
julian.higgins@mrc-bsu.cam.ac.uk

Abstract. We present a revised version of the `metareg` command, which performs meta-analysis regression (meta-regression) on study-level summary data. The major revisions involve improvements to the estimation methods and the addition of an option to use a permutation test to estimate p-values, including an adjustment for multiple testing. We have also made additions to the output, added an option to produce a graph, and included support for the `predict` command. Stata 8.0 or above is required.

Keywords: sbe23_1, meta-regression, meta-analysis, permutation test, multiple testing, metareg

1 Introduction

Meta-analysis regression, or meta-regression, is an extension to standard meta-analysis that investigates the extent to which statistical heterogeneity between results of multiple studies can be related to one or more characteristics of the studies (Thompson and Higgins 2002). Like meta-analysis, meta-regression is usually conducted on study-level summary data, because individual observations from all studies (often referred to as individual patient data in medical applications) are frequently not available.

Sharp (1998) introduced the `metareg` command to perform meta-regression on study-level summary data. In this article, we present a substantially updated and largely rewritten version of `metareg`. The planning and interpretation of meta-regression studies raises substantial statistical issues discussed at length elsewhere (Davey Smith, Egger, and Phillips 1997; Higgins et al. 2002; Thompson and Higgins 2002, 2005). In this article, we will concentrate on the rationale for and the implementation and interpretation of the following new features of `metareg`:

- An improved algorithm for the estimation of the between-study variance, τ^2, by residual (restricted) maximum likelihood (REML)

- A modification to the calculation of standard errors, p-values, and confidence intervals for coefficients suggested by Knapp and Hartung (2003)

- Various enhancements to the output

- An option to produce a graph of the fitted model with a single covariate

- An option to calculate permutation-based p-values, including an adjustment for multiple testing based on the work of Higgins and Thompson (2004)

- Support for many of Stata's postestimation commands, including `predict`

We begin with a brief outline in section 2 of the statistical basis of meta-analysis and meta-regression, and we continue with a summary in section 3 of the relationship of `metareg` to other Stata commands. Section 4 introduces two example datasets that we use to illustrate the discussion of new features in section 5, which constitutes the main body of the article and has subsections corresponding to each of the new features listed above. The final two sections are reference material: Section 7 gives the Stata syntax and full list of options for `metareg` and `predict` after `metareg`, and lists the results saved by the command. Finally, section 7 gives details of the methods and formulas used.

2 Basis of meta-regression

In this section, we outline the statistical basis of random- and fixed-effects meta-regression and their relation to random- and fixed-effects meta-analysis. We will use mathematical formulas for brevity and precision. Less mathematically inclined readers or those who are already familiar with the principles of meta-analysis and meta-regression can skip this section.

We assume that study i of a total of n studies provides an estimate, y_i, of the effect of interest, such as a log odds-ratio, log risk-ratio, or difference in means. Each study also provides a standard error for this estimate, σ_i, which we assume is known, as is common in meta-analysis (although in practice, it will have been estimated from the data in that study). Let us start from the simplest model:

- *Fixed-effects meta-analysis* assumes that there is a single true effect size, θ, so that
$$y_i \sim N(\theta, \sigma_i^2)$$
or equivalently,
$$y_i = \theta + \epsilon_i, \qquad \text{where} \quad \epsilon_i \sim N(0, \sigma_i^2)$$

- *Random-effects meta-analysis* allows the true effects, θ_i, to vary between studies by assuming that they have a normal distribution around a mean effect, θ:
$$y_i \mid \theta_i \sim N(\theta_i, \sigma_i^2), \qquad \text{where} \quad \theta_i \sim N(\theta, \tau^2)$$
So
$$y_i \sim N(\theta, \sigma_i^2 + \tau^2)$$
or equivalently,
$$y_i = \theta + u_i + \epsilon_i, \qquad \text{where} \quad u_i \sim N(0, \tau^2) \text{ and } \epsilon_i \sim N(0, \sigma_i^2)$$

Here τ^2 is the between-study variance and must be estimated from the data.

- *Fixed-effects meta-regression* extends fixed-effects meta-analysis by replacing the mean, θ, with a linear predictor, $\mathbf{x}_i\boldsymbol{\beta}$:

$$y_i \sim N(\theta_i, \sigma_i^2), \qquad \text{where} \quad \theta_i = \mathbf{x}_i\boldsymbol{\beta}$$

or equivalently,

$$y_i = \mathbf{x}_i\boldsymbol{\beta} + \epsilon_i, \qquad \text{where} \quad \epsilon_i \sim N(0, \sigma_i^2)$$

Here $\boldsymbol{\beta}$ is a $k \times 1$ vector of coefficients (including a constant if fitted), and $\mathbf{x}_i$ is a $1 \times k$ vector of covariate values in study i (including a 1 if a constant is fit).

- *Random-effects meta-regression* allows for such residual heterogeneity (between-study variance not explained by the covariates) by assuming that the true effects follow a normal distribution around the linear predictor:

$$y_i \mid \theta_i \sim N(\theta_i, \sigma_i^2), \qquad \text{where} \quad \theta_i \sim N(\mathbf{x}_i\boldsymbol{\beta}, \tau^2)$$

so

$$y_i \sim N(\mathbf{x}_i\boldsymbol{\beta}, \sigma_i^2 + \tau^2)$$

or equivalently,

$$y_i = \mathbf{x}_i\boldsymbol{\beta} + u_i + \epsilon_i, \qquad \text{where} \quad u_i \sim N(0, \tau^2) \text{ and } \epsilon_i \sim N(0, \sigma_i^2)$$

Random-effects meta-regression can be considered either an extension to fixed-effects meta-regression that allows for residual heterogeneity or an extension to random-effects meta-analysis that includes study-level covariates.

Table 1 summarizes the relationships between these models and gives the corresponding Stata commands, which are summarized in the next section.

Table 1. Summary of `metareg` and related Stata commands

	No covariates	With covariate(s)
Fixed-effects model	fixed-effects meta-analysis	fixed-effects meta-regression (not recommended)
	`metan` with `fixedi`, `peto`, or no options	`vwls`
Random-effects model	random-effects meta-analysis	random-effects meta-regression (mixed-effects meta-regression)
	`metan` with `random` or `randomi` options	`metareg`

3 Relation to other Stata commands

Both fixed- and random-effects meta-analysis are available in the user-written package `metan` (Harris et al. 2008). Random-effects meta-analysis can also be performed with `metareg` by not including any covariates (the method-of-moments estimate for between-study variance must be specified to produce identical results to the `metan` command). `metan` can also be used to generate the variables required by `metareg` containing the effect estimate and its standard error for each study from data in various other forms (Harris et al. 2008).

Fixed-effects meta-regression can be fit by weighted least squares by using the official Stata command `vwls` (see [R] **vwls**) with the weights $1/\sigma_i^2$. Fixed-effects meta-regression is not usually recommended, however, because it assumes that all the heterogeneity can be explained by the covariates, and it leads to excessive type I errors when there is residual, or unexplained, heterogeneity (Higgins and Thompson 2004; Thompson and Sharp 1999).

Random-effects meta-regression is closely related to the seldom-used "between-effects" model available in the official Stata command `xtreg` (see [XT] **xtreg**), with studies corresponding to units. Whereas meta-regression assumes that the within-study data have been summarized by an effect estimate, y_i, and its standard error, σ_i, for each study, `xtreg` requires data on individual observations, e.g., individual patient data. Meta-regression is often used on binary outcomes summarized by log odds-ratios or log risk-ratios and their standard errors, whereas `xtreg` is appropriate only for continuous outcomes. `xtreg` also uses different estimators from those available in `metareg`, which are outlined in section 5.1.

4   Background to examples

Our first example is from a meta-analysis of 28 randomized controlled trials of cholesterol-lowering interventions for reducing risk of ischemic heart disease (IHD). The outcome event was death from IHD or nonfatal myocardial infarction. These data are taken from table 1 of Thompson and Sharp (1999). Data from 25 of these trials were also published in Thompson (1993). The measure of effect size is the odds ratio, but statistical analysis is conducted on its natural logarithm, the log odds-ratio, because this has a sampling distribution more closely approximated by a normal distribution. The interventions are varied, with 18 trials of several different drugs, 9 trials of dietary interventions, and 1 trial of a surgical intervention. The eligibility criteria also differed—19 studies recruited only participants without known IHD on entry, 6 recruited only those with IHD, and 3 included those with or without IHD. The reduction in cholesterol varied among trials, as quantified by the difference in mean serum cholesterol concentrations between the treated and control subjects at the end of each trial. Interest focuses on estimating the odds ratio for any given degree of cholesterol reduction (e.g., 1 mmol/L), assuming that any effect on IHD is mediated through the reduction in serum cholesterol. The Stata dataset is named `cholesterol.dta`.

The second example is drawn from a systematic review of 10 randomized controlled trials of exercise as an intervention in the management of depression (Lawlor and Hopker 2001). Here the outcome, severity of depression, was measured on one of two numerical scales, and the measure of effect size was the standardized mean difference. There was considerable between-study heterogeneity in the results of the trials, and the authors considered eight study-level covariates that might explain this heterogeneity. We will focus on the five covariates selected by Higgins and Thompson (2004). The Stata dataset is named `xrcise4deprsn.dta`.

5   New and enhanced features

We now give details of each of the new and enhanced features available in this revision of `metareg`, as listed in section 1. Sections 5.1–5.3 are relevant to all uses of `metareg`. When there is a single continuous covariate, the fitted model can be presented graphically, as shown in section 5.4. Section 5.5 explores a permutation-based approach to calculating p-values, suggested by Higgins and Thompson (2004), who recommended its use when there are few studies and as a way of adjusting for multiple testing when there is more than one covariate of interest. Section 5.6 is intended for more advanced users only; it describes the postestimation facilities available after a `metareg` model has been fit, and it assumes some familiarity with random-effects models, as well as with Stata's graphics commands and postestimation tools.

5.1   Algorithm for REML estimation of τ^2

All algorithms for random-effects meta-regression first estimate the between-study variance, τ^2, and then estimate the coefficients, β, by weighted least squares by using the weights $1/(\sigma_i^2 + \tau^2)$, where σ_i^2 is the standard error of the estimated effect in study i. The default algorithm in `metareg` is REML, as advocated by Thompson and Sharp (1999).

The algorithm for REML estimation has been improved in this update of `metareg`. The original version used an iterative algorithm (Morris 1983) that was not guaranteed to converge and was only an approximation when the within-study standard errors varied. The original version of `metareg` sometimes misleadingly reported an estimate of $\hat{\tau}^2 = 0$ when the algorithm was in fact diverging (for example, with the `cholesterol` data). This revised version of `metareg` instead directly maximizes the residual (restricted) log likelihood by using Stata's robust and well-tested `ml` command, avoiding the approximations and convergence problems of the previous method.

We decided not to implement the standard maximum likelihood (ML) estimator in this updated version of `metareg`. (To ensure all do-files written for the original version of `metareg` continue to work, however, the code of the original program is included in this package so that a request for the ML estimator can be handled by calling the original code.) Both REML and ML are iterative methods. Unlike REML, however, ML does not account for the degrees of freedom used in estimating the fixed effects. This can make a particular difference in meta-regression because the number of observations (studies) is often small. As a result, the ML estimate of τ^2 is often biased downward, leading to underestimated standard errors and anticonservative inference (Thompson and Sharp 1999; Sidik and Jonkman 2007).

Further details of the methods for the estimation of τ^2 are given in section 7.1.

5.2 Knapp–Hartung variance estimator and associated t test

Knapp and Hartung (2003) introduced a novel estimator for the variances of the effect estimates in meta-regression. Their variance estimator amounts to calculating a quadratic form, q, and multiplying the usual variance estimates by q if $q > 1$. This estimator should be used with a t distribution when calculating p-values and confidence intervals. They found this procedure to have much more appropriate false-positive rates than the standard approach, a finding confirmed by Higgins and Thompson (2004) in more extensive simulations.

We therefore recommend this variance estimator and have made it the default in `metareg`. It is particularly suitable for estimation of standard errors and confidence intervals. However, it can be unreasonably conservative (false-positive rates below the nominal level) when the number of studies is particularly small, further reducing the already limited power. When there are few studies, the permutation test detailed in section 5.5 below has the potential to provide a better, though more computationally intensive, method for calculating p-values.

5.3   Enhancements to the output

The following additions have been made to the output of `metareg` that is displayed above the coefficient table:

- A measure of the percentage of the residual variation that is attributable to between-study heterogeneity (I^2_{res})

- The proportion of between-study variance explained by the covariates (a type of adjusted R^2 statistic)

- An overall test of all the covariates in the random-effects model

The iteration log is no longer displayed by default.

We will illustrate these additions by using the output of `metareg` in the simplest situation where a single continuous covariate is fit, using the `cholesterol` data as an example:

```
. use cholesterol
(Serum cholesterol reduction & IHD)

. metareg logor cholreduc, wsse(selogor)

Meta-regression                                      Number of obs   =       28
REML estimate of between-study variance              tau2            =   .0097
% residual variation attributable to heterogeneity   I-squared_res   =  31.34%
Proportion of between-study variance explained       Adj R-squared   =  69.02%
With Knapp-Hartung modification
```

logor	Coef.	Std. Err.	t	P>|t|	[95% Conf.	Interval]
cholreduc	-.5056849	.1834858	-2.76	0.011	-.8828453	-.1285244
_cons	.1467225	.1374629	1.07	0.296	-.1358367	.4292816

Residual heterogeneity of the fixed-effects model

The residual heterogeneity statistic is the weighted sum of squares of the residuals from the fixed-effects meta-regression model and is a generalization of Cochran's Q from meta-analysis to meta-regression. To distinguish it from the total heterogeneity statistic Q that would be obtained from ordinary meta-analysis, i.e., without fitting any covariates, we will denote it by Q_{res} (Lipsey and Wilson [2001] denote the same statistic by Q_E). A test of the null hypothesis of no residual (unexplained) heterogeneity can be obtained by comparing Q_{res} to a χ^2 distribution with $n - k$ degrees of freedom. However, it is often more useful to quantify heterogeneity than to test for it (Higgins et al. 2003): The proportion of residual between-study variation due to heterogeneity, as opposed to sampling variability, is calculated as $I^2_{\text{res}} = \max[0, \{Q_{\text{res}} - (n - k)\}/Q_{\text{res}}]$, an obvious extension to the I^2 measure in meta-analysis (Higgins et al. 2003).

From the value of I^2_{res} in the output above, 31% of the residual variation is due to heterogeneity, with the other 69% attributable to within-study sampling variability.

Adjusted R^2

The proportion of between-study variance explained by the covariates can be calculated by comparing the estimated between-study variance, $\hat{\tau}^2$, with its value when no covariates are fit, $\hat{\tau}_0^2$. Adjusted R^2 is the relative reduction in the between-study variance, $R_{\text{adj}}^2 = (\hat{\tau}_0^2 - \hat{\tau}^2)/\hat{\tau}_0^2$. It is possible for this to be negative if the covariates explain less of the heterogeneity than would be expected by chance, but the same is true for adjusted R^2 in ordinary linear regression. It may be more common in meta-regression because the number of studies is often small.

In the above example, 69% of the between-study variance is explained by the covariate `cholreduc`, and the remaining between-study variance appears small at 0.0097. (It is coincidence that the figure of 69% also appears in the preceding subsection.)

Joint test for all covariates

When more than one covariate is fit, `metareg` reports a test of the null hypothesis that the coefficients of the covariates are all zero, obtained from a multiparameter Wald test by using Stata's `test` command (see [R] **test**). The test statistic is compared to the appropriate F distribution if the default Knapp–Hartung adjustment is used. If `metareg`'s z option is used to specify the use of conventional variance estimates and tests for the effect estimates, a χ^2 distribution is used for the joint test. To simplify the output, this test is not displayed when only a single covariate is fit because it would give an identical p-value to the one displayed for the covariate in the regression table.

This gives one way of controlling the risk of false-positive findings when performing meta-regression with multiple covariates: we can use the overall model p-value to assess if there is evidence for an association of *any* of the covariates with the outcome. However, when a small p-value indicates that there is such evidence, it becomes harder to decide which, and how many, of the covariates there is good evidence for. Another method of dealing with this multiplicity issue that may help overcome this problem, though at the expense of longer computation time, is given in section 5.5 below.

(*Continued on next page*)

▷ **Example**

We illustrate this joint test by using all five covariates available in the data on exercise for depression:

```
. use xrcise4deprsn
(Exercise for depression)

. metareg smd abstract-phd, wsse(sesmd)

Meta-regression                                          Number of obs  =        10
REML estimate of between-study variance                  tau2           =         0
% residual variation attributable to heterogeneity       I-squared_res  =    0.00%
Proportion of between-study variance explained           Adj R-squared  = 100.00%
Simultaneous test for all covariates                     Model F(5,4)   =      6.57
With Knapp-Hartung modification                          Prob > F       =    0.0460
```

smd	Coef.	Std. Err.	t	P>|t|	[95% Conf. Interval]	
abstract	-1.33993	.3892562	-3.44	0.026	-2.420678	-.2591814
duration	.1567629	.0616404	2.54	0.064	-.0143784	.3279041
itt	.4611682	.3883635	1.19	0.301	-.6171018	1.539438
alloc	-.4063866	.3503447	-1.16	0.311	-1.379099	.5663263
phd	-.0138045	.440595	-0.03	0.977	-1.237092	1.209483
_cons	-2.07241	.5683944	-3.65	0.022	-3.650526	-.4942942

Here $\hat{\tau}^2$ is zero, and it follows that $I^2_{\text{res}} = 0\%$ and $R^2_{\text{adj}} = 100\%$. The joint test for all five covariates gives a p-value of 0.046, indicating some evidence for an association of at least one of the covariates with the size of the treatment effect.

◁

5.4   Graph of the fitted model

When a single continuous covariate is fit, one common way to present the fitted model, sometimes referred to as a "bubble plot", is to graph the fitted regression line together with circles representing the estimates from each study, sized according to the precision of each estimate (the inverse of its within-study variance, σ_i^2). The graph option to metareg gives an easy way to produce such a plot, as illustrated in figure 1 for the cholesterol data.

```
. use cholesterol
(Serum cholesterol reduction & IHD)
. metareg logor cholreduc, wsse(selogor) graph
   (output omitted)
```

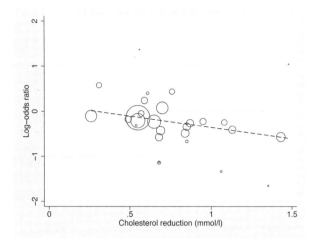

Figure 1. "Bubble plot" with fitted meta-regression line

An additional option, `randomsize`, is provided for those who prefer the size of the circles to depend on the weight of the study in the fitted random-effects meta-regression model (the inverse of its total variance, $\sigma_i^2 + \hat{\tau}^2$). This makes only a slight difference to the example above because the estimated between-study variance, $\hat{\tau}^2$, is small; in general, though, it will give circles that vary less in size.

Those wishing to further customize the plot can use the `predict` command to generate fitted values followed by a `graph twoway` command (see section 5.6).

5.5   Permutation test

Higgins and Thompson (2004) proposed using a permutation test approach to calculating p-values in meta-regression. Permutation tests provide a nonparametric way of simulating data under the null hypothesis (see, e.g., Manly [2006]). Calculation of exact permutation p-values would be feasible when there are few studies by enumeration of all possible permutations, but for simplicity, we have implemented a permutation test based on Monte Carlo simulation, i.e., based on *random* permutations.

The algorithm is similar to other applications of permutation methods, and it is implemented with Stata's `permute` command (see [R] **permute**). The covariates are randomly reallocated to the outcomes many times, and a t statistic is calculated each time. The true p-value for the relationship between a given covariate and the response is computed by counting the number of times these t statistics are greater than or equal to the observed t statistic. When multiple covariates are included in the meta-regression, the covariate values for a given study are kept together to preserve and account for their correlation structure. In meta-regression, unlike other regressions, the outcome consists of both the effect size and its standard error, and these must be kept together. This small complication makes it impossible to use `permute` on `metareg` directly from the command line when there are multiple covariates, so we have written a `permute()` option for `metareg`. This option also implements the following extension, which adjusts *p*-values for multiple tests when there are several covariates.

Multiplicity adjustment

When several covariates are used in meta-regression, either in several separate univariable meta-regressions or in one multiple meta-regression, there is an increased chance of at least one false-positive finding (type I error). The statistics obtained from the random permutations can be used to adjust for such multiple testing by comparing the observed *t* statistic for every covariate with the largest *t* statistic for any covariate in each random permutation. The proportion of times that the former equals or exceeds the latter gives the probability of observing a *t* statistic for any covariate as extreme or more extreme than that observed for a particular covariate, under the complete null hypothesis that all the regression coefficients are zero.

The number of random permutations must be specified—there is deliberately no default. We suggest that a small number (e.g., 100) be specified initially to check that the command is working as expected. The number should then be increased to at least 1,000, but 5,000 or 20,000 permutations may be necessary for sufficient precision (Manly 2006; Westfall and Young 1993). Because the `permute()` option uses Stata's random-number generator, the `set seed` command should be used first if replicability of results is desired. When the `permute()` option is specified, the defaults are to use the method-of-moments estimate of τ^2 for reasons of speed and to not use the Knapp–Hartung modification to the standard errors.

By default, `permute()` performs multivariable meta-regression; i.e., all the covariates are entered into a single model in each permutation.

▷ **Example**

We illustrate the use of the `permute()` option by using the data on exercise for depression.

```
. use xrcise4deprsn
(Exercise for depression)

. set seed 15160401

. metareg smd abstract-phd, wsse(sesmd) permute(20000)

Monte Carlo permutation test for meta-regression

Moment-based estimate of between-study variance
Without Knapp & Hartung modification to standard errors

P-values unadjusted and adjusted for multiple testing
                 Number of obs =        10
                 Permutations   =     20000

                           P
      smd │ Unadjusted   Adjusted
──────────┼──────────────────────
 abstract │      0.023      0.089
 duration │      0.056      0.201
      itt │      0.311      0.721
    alloc │      0.313      0.736
      phd │      0.978      1.000
──────────┼

largest Monte Carlo SE(P) = 0.0033
WARNING:
Monte Carlo methods use random numbers, so results may differ between runs.
Ensure you specify enough permutations to obtain the desired precision.
```

The first column of the results table gives permutation p-values without an adjustment for multiplicity. The results are in good agreement with the p-values obtained in section 5.3 without using the permutation option but with the Knapp–Hartung modification. The second column gives p-values adjusted for multiplicity. We see that all the p-values are increased. After adjusting for multiple testing, there remains some weak evidence that results of studies published as an abstract differ on average from results of studies published as a full article. The adjusted p-value of 0.089 gives the probability under the complete null hypothesis (that all regression coefficients are zero) of a t statistic for *any* of the five covariates as extreme or more extreme as that observed for the covariate `abstract`. As Higgins and Thompson (2004) suggest, this can be interpreted as describing the degree of "surprise" one might have about the observed result for this covariate, considering that five covariates are being examined. This is less conservative than the Bonferroni adjusted p-value of $0.0235 \times 5 = 0.1175$.

The output also gives the largest Monte Carlo standard error of the calculated p-values as an indication of the degree of precision obtained by the specified number of random permutations. Standard errors and "exact" confidence intervals for each of the p-values can be obtained by using the `detail` suboption. (These can always be calculated afterward by using the `cii` command if this option was not specified.)

◁

❏ Technical note

Higgins and Thompson (2004) originally proposed a slightly different permutation-based multiplicity adjustment: it compared the ith largest t statistic observed (for the "*i*th most significant" covariate) with the *i*th largest t statistic in each random permutation. This adjustment was implemented in a revised version of `metareg` released previously on the Statistical Software Components archive. This adjustment has been found to be hard to interpret in practice, however, because for the second most significant covariate it effectively gives a *joint* test of the two covariates with the largest two observed t statistics (and similarly for third and subsequent covariates if more than two covariates are supplied). The resulting multiplicity-adjusted p-value can turn out to be either larger *or smaller* than the unadjusted p-value, which can appear counter-intuitive.

For this release of `metareg`, we have therefore chosen to implement a different permutation-based algorithm for multiplicity adjustment based on the one-step "maxT" method of Westfall and Young (1993). This adjustment compares the t statistic for *every* covariate with the largest t statistic in each random permutation. The resulting multiplicity-adjusted p-values are always as large as or (usually) larger than the unadjusted p-values. This procedure ensures weak control of the familywise error rate, defined as the probability that at least one null hypothesis is rejected when *all* the null hypotheses are true (Shaffer 1995). It does not guarantee strong control of the familywise error rate, however; i.e., when one or more null hypotheses are false, it does not guarantee control of the proportion of the remaining true null hypotheses that are incorrectly rejected, though such strong control should be achieved asymptotically as the number of studies increases (Westfall and Young 1993; Shaffer 1995).

The false discovery rate (Benjamini and Hochberg 1995) and related procedures (Newson and the ALSPAC Study Team 2003; Storey, Taylor, and Siegmund 2004; Wacholder et al. 2004) have been suggested as an alternative method of multiplicity adjustment, but we have chosen not to implement such procedures in `metareg`. Such procedures are always either step-up or (more rarely) step-down algorithms. Although stepwise algorithms are suitable for hypothesis testing and often give greater power, the resulting adjusted p-values cannot be interpreted as giving the strength of evidence against the null hypothesis, the interpretation increasingly advocated in medicine and epidemiology (Sterne and Davey Smith 2001). In particular, stepwise methods may assign equal adjusted p-values to covariates with different unadjusted p-values.

❏

Suboptions to permute()

The `permute()` option can also be used to perform a set of single-variable meta-regressions at each permutation by adding the `univariable` suboption. This suboption reports permutation-based p-values for fitting a separate model for each covariate rather than including all the covariates in a multiple regression model. With several covariates, the execution time may be considerably longer than for multivariable meta-regression.

▷ **Example**

We add the `univariable` suboption to the previous example but reduce the number of permutations to cut down the computation time:

```
. metareg smd abstract-phd, wsse(sesmd) permute(5000, univariable)
Monte Carlo permutation test for single covariate meta-regressions
Moment-based estimate of between-study variance
Without Knapp & Hartung modification to standard errors
P-values unadjusted and adjusted for multiple testing
              Number of obs =        10
              Permutations  =      5000

                       P
      smd │ Unadjusted   Adjusted
──────────┼──────────────────────
 abstract │    0.021       0.043
 duration │    0.030       0.115
      itt │    0.384       0.946
    alloc │    0.330       0.861
      phd │    0.715       0.999
──────────┴──────────────────────

largest Monte Carlo SE(P) = 0.0069
WARNING:
Monte Carlo methods use random numbers, so results may differ between runs.
Ensure you specify enough permutations to obtain the desired precision.
```

In these results, unlike those from the previous example, each covariate is fit in a separate model and so is not adjusted for the other covariates. The p-values do not differ greatly in this example, however.

◁

There is also a `joint()` suboption that requests a permutation p-value for a joint test of the variables specified. This can be particularly useful if a set of indicator variables is used to model a categorical covariate.

A joint test of covariates can be obtained without using a permutation approach by instead using the `test` or `testparm` (see [R] **test**) command after `metareg`.

A p-value for the joint test is not included in the multiplicity-adjustment procedure because the two are neither technically nor philosophically compatible.

▷ Example

We return to the `cholesterol` data, in which the `ihdentry` variable is a categorical covariate with three categories indicating whether the study included participants with known IHD on entry to the study, without known IHD, or both:

```
. use cholesterol
(Serum cholesterol reduction & IHD)

. tab ihdentry, gen(ihd)
    Ischaemic heart │
    disease on entry │      Freq.     Percent        Cum.
────────────────────┼───────────────────────────────────
   Without known IHD │          6       21.43       21.43
            With IHD │         19       67.86       89.29
  With or without IHD │          3       10.71      100.00
────────────────────┼───────────────────────────────────
               Total │         28      100.00
```

```
. metareg logor cholreduc ihd2 ihd3, wsse(selogor)
> permute(5000, joint(ihd2 ihd3))

Monte Carlo permutation test for meta-regression

Moment-based estimate of between-study variance
Without Knapp & Hartung modification to standard errors
joint1 : ihd2 ihd3

P-values unadjusted and adjusted for multiple testing

                    Number of obs =        28
                    Permutations  =      5000
```

		P
logor	Unadjusted	Adjusted
cholreduc	0.009	0.028
ihd2	0.611	0.933
ihd3	0.907	0.999
joint1	0.883	

```
largest Monte Carlo SE(P) = 0.0069
WARNING:
Monte Carlo methods use random numbers, so results may differ between runs.
Ensure you specify enough permutations to obtain the desired precision.
```

The p-value of 0.883 for the joint test of ihd2 and ihd3 indicates that there is very little evidence that the log odds-ratio differs among these three categories of studies, after adjusting for the degree of cholesterol reduction achieved in each study.

◁

5.6 Postestimation tools for metareg

metareg is programmed as a Stata estimation command and so supports most of Stata's postestimation commands (except when the **permute()** option is used). (One deliberate exception is lrtest, which is not appropriate after metareg because the REML log likelihood cannot be used to compare models with different fixed effects, while the method of moments does not give a likelihood.)

Several quantities can be obtained by using **predict** after **metareg**, including fitted values and predicted random effects (empirical Bayes estimates). These can be useful for producing graphs of the fitted model and for model checking. Details of the syntax and options are given in sections 6.4 and 6.5, and section 7.4 contains the formulas used.

We now illustrate the use of some of the quantities available from **predict** in a graph. Using the exercise for depression data, we conduct a meta-regression of the standardized mean difference on the single covariate duration that describes the duration of follow-up in each study. Figure 2 shows the fitted line and the estimates from the separate studies that would be produced by the **graph** option to **metareg**, and it also includes the empirical Bayes estimates and shaded bands showing both confidence and prediction intervals (we would not recommend including all these features on a single graph in practice). It was produced by the following commands:

```
. use xrcise4deprsn, clear
(Exercise for depression)

. metareg smd duration, wsse(sesmd)

Meta-regression                                      Number of obs  =      10
REML estimate of between-study variance              tau2           =   .2019
% residual variation attributable to heterogeneity   I-squared_res  =  55.83%
Proportion of between-study variance explained       Adj R-squared  =  55.16%
With Knapp-Hartung modification
```

smd	Coef.	Std. Err.	t	P>|t|	[95% Conf. Interval]	
duration	.2097633	.0802611	2.61	0.031	.0246808	.3948457
_cons	-2.907511	.7339255	-3.96	0.004	-4.599946	-1.215076

```
. predict fit
(option xb assumed; fitted values)

. predict stdp, stdp

. predict stdf, stdf

. predict xbu, xbu

. local t = invttail(e(df_r)-1, 0.025)

. gen confl = fit - `t'*stdp

. gen confu = fit + `t'*stdp

. gen predl = fit - `t'*stdf

. gen predu = fit + `t'*stdf

. sort duration

. twoway rarea predl predu duration || rarea confl confu duration
> || line fit duration
> || scatter smd duration [aw=1/sesmd^2], msymbol(Oh)
> || scatter xbu duration, msymbol(t)
> ||, legend(label(1 "Prediction interval") label(2 "Confidence interval")
> cols(1))
```

(Continued on next page)

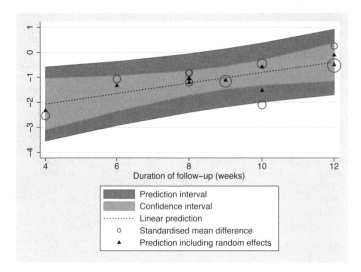

Figure 2. Confidence and prediction intervals and empirical Bayes estimates

The `stdp` option to `predict` gives the standard error of the fitted values excluding the random effects, commonly referred to as the standard error of the prediction. This standard error is used to draw a pointwise confidence interval, shown in light gray in figure 2, around the fitted line, illustrating our uncertainty about the position of the line. The `stdf` option to `predict` gives the standard deviation of the predicted distribution of the true value of the outcome in a future study with a given value of the covariate(s), commonly referred to as the standard error of the forecast. This standard error is used to draw a prediction interval, shown in dark gray in figure 2, around the fitted line, illustrating our uncertainty about the true effect we would predict in a future study with a known duration of follow-up. The prediction band will be wider than the confidence band unless $\tau^2 = 0$. The use of a t distribution in generating the intervals is an approximation, and opinions differ over the most appropriate degrees of freedom; we use $n - k - 1$ here to be consistent with the $n - 2$ used by Higgins, Thompson, and Spiegelhalter (2009) for confidence and prediction intervals in meta-analysis, where $k = 1$. The `xbu` option to `predict` gives the empirical Bayes estimates (predictions including random effects), shown as triangles in figure 2. These are our best estimates of the true effect in each study, assuming the fitted model is correct. If I_{res}^2 is small, the empirical Bayes estimates will tend to lie well inside the prediction interval; if $\tau^2 = 0$, implying $I_{\mathrm{res}}^2 = 0$, they will all lie on the fitted line.

The statistics available from `predict` can also be useful for model checking and checking for outliers and influential studies. This checking is best done graphically. One possibility is a normal probability plot of the standardized predicted random effects (equivalently, standardized empirical Bayes residuals, or standardized shrunken residuals; see figure 3). This probability plot can be used to check the assumption of normality of the random effects, although because this assumption has been used in generating the predictions, only gross deviations are likely to be detected. Perhaps more usefully, the probability plot can be used to detect outliers:

```
. use cholesterol, clear
(Serum cholesterol reduction & IHD)
. qui metareg logor cholreduc, wsse(selogor)
. capture drop usta
. predict usta, ustandard
. qnorm usta, mlabel(id)
```

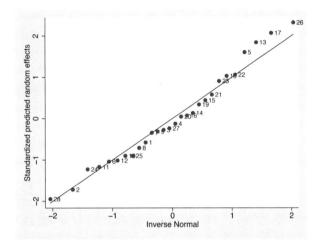

Figure 3. Normal probability plot of standardized shrunken residuals

Figure 3 suggests that the assumption of normal random effects is adequate, and there are no notable outliers because the largest standardized shrunken residual is only slightly over 2.

Other plots useful for model checking and identifying influential points in conventional linear regression may also be useful for meta-regression, for example, leverage–residual (L–R) plots, or plots of residuals versus either fitted values or a predictor; see [R] **regress postestimation** for further details of these and other plots (the various plot commands given there will not work after `metareg`, but it should be fairly straightforward to use `predict` followed by the appropriate `graph twoway` command to produce similar plots).

6 Syntax, options, and saved results

6.1 Syntax

The syntax of `metareg` has been revised somewhat from that of the original version (Sharp 1998). The original syntax should continue to work, but it is not documented here. ML estimation of τ^2 is not supported by the updated `metareg` program, but if the old `bsest(ml)` option is used, the new program simply calls the original version, which is incorporated within the updated `metareg.ado` file.

`metareg` *depvar* [*indepvars*] [*if*] [*in*] `wsse`(*varname*) [, e̲form graph
    randomsize no̲constant mm reml eb k̲napphartung z tau2test l̲evel(#)
    permute(# [, u̲nivariable d̲etail j̲oint(*varlist1* [| *varlist2* ...])])]) log
    maximize_options]

`by` can be used with `metareg`; see [D] **by**.

6.2 Options

`wsse`(*varname*) specifies the variable containing σ_i, the standard error of *depvar*, within each study. All values of *varname* must be greater than zero. `wsse()` is required.

`eform` indicates to output the exponentiated form of the coefficients and to suppress reporting of the constant. This option may be useful when *depvar* is the logarithm of a ratio measure, such as a log odds-ratio or a log risk-ratio.

`graph` requests a line graph of fitted values plotted against the first covariate in *indepvars*, together with the estimates from each study represented by circles. By default, the circle sizes depend on the precision of each estimate (the inverse of its within-study variance), which is the weight given to each study in the fixed-effects model.

`randomsize` is for use with the `graph` option. It specifies that the size of the circles will depend on the weights in the random-effects model rather than the precision of each estimate. These random-effects weights depend on the estimate of τ^2.

The remaining options will mainly be of interest to more advanced users:

`noconstant` suppresses the constant term (intercept). This is rarely appropriate in meta-regression. *Note*: It might seem tempting to use the `noconstant` option in the cholesterol example to force the regression line through the origin, on the reasoning that an intervention that has no effect on cholesterol should have no effect on the odds of IHD. We would advise against using this option, however, both here and in most other circumstances. Using it here involves the assumption that the effect of the intervention on IHD is mediated entirely by cholesterol reduction. It also would not allow for measurement error in cholesterol reduction, which, through attenuation by errors (regression dilution bias), could lead to a nonzero intercept even when a zero intercept would be expected.

The `mm`, `reml`, and `eb` options are alternatives that specify the method of estimation of the additive (between-study) component of variance τ^2:

mm specifies the use of method of moments to estimate the additive (between-study) component of variance τ^2; this is a generalization of the DerSimonian and Laird (1986) method commonly used for random-effects meta-analysis. For speed, this is the default when the **permute()** option is specified, because it is the only noniterative method.

reml specifies the use of REML to estimate the additive (between-study) component of variance τ^2. This is the default unless the **permute()** option is specified. This revised version uses Stata's ML facilities to maximize the REML log likelihood. It will therefore not give identical results to the previous version of **metareg**, which used an approximate iterative method.

eb specifies the use of the "empirical Bayes" method to estimate τ^2 (Morris 1983).

knapphartung makes a modification to the variance of the estimated coefficients suggested by Knapp and Hartung (2003) and supported by Higgins and Thompson (2004), accompanied by the use of a t distribution in place of the standard normal distribution when calculating p-values and confidence intervals. This is the default unless the **permute()** option is specified.

z requests that the **knapphartung** modification not be applied and that the standard normal distribution be used to calculate p-values and confidence intervals. This is the default when the **permute()** option is specified with a fixed-effects model.

tau2test adds to the output two tests of $\tau^2 = 0$. The first is based on the residual heterogeneity statistic, Q_{res}. The second (not available if the **mm** option is also specified) is a likelihood-ratio test based on the REML log likelihood. These are two tests of the same null hypothesis (the fixed-effects model with $\tau^2 = 0$), but the alternative hypotheses are different, as are the distributions of the test statistics under the null, so close agreement of the two tests is not guaranteed. Both tests are typically of little interest because it is more helpful to quantify heterogeneity than to test for it (see section 5.3).

level(#) specifies the confidence level, as a percentage, for confidence intervals. The default is **level(95)** or as set by **set level**; see [U] **20.7 Specifying the width of confidence intervals**.

permute(...) calculates p-values by using a Monte Carlo permutation test. See section 6.3 below for more information about this option.

log requests the display of the iteration log during estimation of τ^2. This is ignored if the **mm** option is specified, because this uses a noniterative method.

maximize_options are ignored unless estimation of τ^2 is by REML. These options control the maximization process; see [R] **maximize**. They are ignored if the **mm** option is specified. You should never need to specify them; they are supported only in case problems in the REML estimation of τ^2 are ever reported or suspected.

6.3 Option for permutation test

The permute() option calculates p-values by using a Monte Carlo permutation test, as recommended by Higgins and Thompson (2004). To address multiple testing, permute() also calculates p-values for the most- to least-significant covariates, as the same authors also recommend.

The syntax of permute() is

permute(# [, univariable detail joint(*varlist1* [| *varlist2* ...])])

where # is required and specifies the number of random permutations to perform. Larger values give more precise p-values but take longer.

There are three suboptions:

univariable indicates that p-values should be calculated for a series of single covariate meta-regressions of each covariate in *varlist* separately, instead of a multiple meta-regression of all covariates in *varlist* simultaneously.

detail requests lengthier output in the format given by [R] **permute**.

joint(*varlist1* [| *varlist2* ...]) specifies that a permutation p-value should also be computed for a joint test of the variables in each varlist.

The eform, level(), and z options have no effect when the permute() option is specified.

6.4 Syntax of predict

The syntax of predict (see [R] **predict**) following metareg is

predict [*type*] *newvar* [*if*] [*in*] [, *statistic*]

statistic	description
xb	fitted values; the default
stdp	standard error of the prediction
stdf	standard error of the forecast
u	predicted random effects
ustandard	standardized predicted random effects
xbu	prediction including random effects
stdxbu	standard error of xbu
hat	leverage (diagonal elements of hat matrix)

These statistics are available both in and out of sample; type predict ... if e(sample) ... if wanted only for the estimation sample.

6.5   Options for predict

xb, the default, calculates the linear prediction, $\mathbf{x}_i\mathbf{b}$, that is, the fitted values *excluding* the random effects.

stdp calculates the standard error of the prediction (the standard error of the fitted values excluding the random effects).

stdf calculates the standard error of the forecast. This gives the standard deviation of the predicted distribution of the true value of *depvar* in a future study, with the covariates given by *varlist*. $\mathtt{stdf}^2 = \mathtt{stdp}^2 + \widehat{\tau}^2$.

u calculates the predicted random effects, u_i. These are the best linear unbiased predictions of the random effects, also known as the empirical Bayes (or posterior mean) estimates of the random effects, or as shrunken residuals.

ustandard calculates the standardized predicted random effects, i.e., the predicted random effects, u_i, divided by their (unconditional) standard errors. These may be useful for diagnostics and model checking.

xbu calculates the prediction *including* the random effects, $\mathbf{x}_i\mathbf{b} + u_i$, also known as the empirical Bayes estimates of the effects for each study.

stdxbu calculates the standard error of the prediction including random effects.

hat calculates the leverages (the diagonal elements of the projection hat matrix).

6.6   Saved results

When the permute() option is not specified, metareg saves the following in e():

Scalars

e(N)	number of observations	e(tau2)	estimate of τ^2
e(df_m)	model degrees of freedom	e(Q)	Cochran's Q
e(df_Q)	degrees of freedom for test of $Q = 0$	e(I2)	I-squared
		e(q_KH)	Knapp–Hartung variance modification factor
e(df_r)	residual degrees of freedom (if t tests used)	e(remll_c)	REML log likelihood, comparison model
e(remll)	REML log likelihood	e(tau2_0)	τ^2, constant-only model
e(chi2_c)	χ^2 for comparison test	e(chi2)	model χ^2
e(F)	model F statistic		

Macros

e(cmd)	metareg	e(depvar)	name of dependent variable
e(predict)	program used to implement predict	e(method)	REML, Method of moments, or Empirical Bayes
e(wsse)	name of wsse() variable	e(properties)	b V

Matrices

e(b)	coefficient vector	e(V)	variance–covariance matrix of estimators

Functions

e(sample)	marks estimation sample

`metareg, permute()` saves the following in `r()`:

Scalars
    `r(N)`   number of observations

Matrices
    `r(b)`   observed t statistics, T_{obs} `r(p)`   observed proportions
    `r(c)`   count when $|T| \geq |T_{\mathrm{obs}}|$ `r(reps)`   number of nonmissing results

7 Methods and formulas

The residual heterogeneity statistic, Q_{res}, is the residual weighted sum of squares from the fixed-effects model and is the same as the goodness-of-fit statistic computed by `vwls`:

$$Q_{\mathrm{res}} = \sum_i \left(\frac{y_i - \mathbf{x}_i \widehat{\boldsymbol{\beta}}}{\sigma_i} \right)^2$$

The proportion of residual variation due to heterogeneity is

$$I^2 = \max \left\{ \frac{Q_{\mathrm{res}} - (n-k)}{Q_{\mathrm{res}}},\ 0 \right\}$$

The proportion of the between-study variance explained by the covariates (adjusted R-squared) is $R_a^2 = (\widehat{\tau}_0^2 - \widehat{\tau}^2)/\widehat{\tau}_0^2$, where $\widehat{\tau}^2$ and $\widehat{\tau}_0^2$ are the estimates of the between-study variance in models with and without the covariates, respectively.

7.1 Estimation of τ^2

Several different algorithms have been proposed for estimation of the between-study variance, τ^2, in meta-analysis (Sidik and Jonkman 2007) and meta-regression (Thompson and Sharp 1999). Three algorithms are available in this version of `metareg`. In each case, if the estimated value of τ^2 is negative, it is set to zero.

Method of moments is the only noniterative method, so it has the advantages of speed and robustness. It is the natural extension of the DerSimonian and Laird (1986) estimate commonly used in random-effects meta-analysis. The method-of-moments estimate of τ^2 is obtained by equating the observed value of Q_{res} to its expected value under the random-effects model, giving (DuMouchel and Harris 1983, eq. 3.12)

$$\widehat{\tau}_{\mathrm{MM}}^2 = \frac{Q_{\mathrm{res}} - (n+k)}{\sum_i \{1/\sigma_i^2(1-h_i)\}}$$

Here h_i is the ith diagonal element of the hat matrix $\mathbf{X}(\mathbf{X}'\mathbf{V}_0^{-1}\mathbf{X})^{-1}\mathbf{X}\mathbf{V}_0^{-1}$, where $\mathbf{V}_0 = \mathrm{diag}(\sigma_1^2, \sigma_2^2, \ldots, \sigma_n^2)$.

The iterative methods below use $\widehat{\tau}_{MM}^2$ as a starting value (this is a change from the original version of `metareg` (Sharp 1998), which used zero as a starting value).

REML estimation of τ^2 is based on maximization of the residual (or restricted) log likelihood,

$$L_R(\tau^2) = -\frac{1}{2} \sum_i \left\{ \log(\sigma_i^2 + \tau^2) + \frac{(y_i - \mathbf{x}_i \widehat{\boldsymbol{\beta}})^2}{\sigma_i^2 + \tau^2} \right\} - \frac{1}{2} \log |\mathbf{X}' \mathbf{V}^{-1} \mathbf{X}|$$

where $\mathbf{V} = \mathrm{diag}(\sigma_1^2 + \tau^2, \sigma_2^2 + \tau^2, \ldots, \sigma_n^2 + \tau^2)$ and $\widehat{\boldsymbol{\beta}} = (\mathbf{X}' \mathbf{V}^{-1} \mathbf{X})^{-1} \mathbf{X}' \mathbf{V}^{-1} \mathbf{y}$ (Harville 1977). This log likelihood is maximized by Stata's `ml` command, using the `d0` method, which calculates all derivatives numerically.

The "empirical Bayes" estimator of τ^2 is so named because of its introduction in an article on empirical Bayes inference by Morris (1983), although as he states, any approximately unbiased estimate of τ^2 could be used in such a setting. Thompson and Sharp (1999) found it to give substantially larger estimates of τ^2 than other methods. Others suggest it performs well in simulations based on 2×2 tables (Berkey et al. 1995; Sidik and Jonkman 2007), although this may be due to overestimation of the within-study standard errors in small studies by the conventional (Woolf) estimate rather than the properties of the empirical Bayes method itself (Sutton and Higgins 2008). It can also be considered to be a method-of-moments estimator, formed by equating the weighted sum of squares of the residuals from the random-effects model to its expected value (Knapp and Hartung 2003). It is found by iterating the following equation (Morris 1983; Berkey et al. 1995):

$$\widehat{\tau}_{\mathrm{EB}}^2 = \frac{n/(n-k) \sum_i \left\{ (y_i - \mathbf{x}_i \widehat{\boldsymbol{\beta}})^2 / (\sigma_i^2 + \widehat{\tau}_{\mathrm{EB}}^2) - \sigma_i^2 \right\}}{\sum_i (\sigma_i^2 + \widehat{\tau}_{\mathrm{EB}}^2)^{-1}}$$

At each iteration, $\widehat{\boldsymbol{\beta}}$ must be reestimated using a weighted least-squares regression of $\mathbf{y}$ on $\mathbf{X}$ with the weights $1/(\sigma_i^2 + \widehat{\tau}_{\mathrm{EB}}^2)$.

7.2   Estimation of β

Once τ^2 has been estimated by one of the methods above, the estimated coefficients, $\widehat{\boldsymbol{\beta}}$, are obtained by a weighted least-squares regression of $\mathbf{y}$ on $\mathbf{X}$ with the weights $1/(\sigma_i^2 + \widehat{\tau}^2)$. The conventional estimate of the variance–covariance matrix of the coefficients is $(\mathbf{X}' \widehat{\mathbf{V}}^{-1} \mathbf{X})^{-1}$, where $\widehat{\mathbf{V}} = \mathrm{diag}(\sigma_1^2 + \widehat{\tau}^2, \sigma_2^2 + \widehat{\tau}^2, \ldots, \sigma_n^2 + \widehat{\tau}^2)$.

7.3   Knapp–Hartung variance modification

Knapp and Hartung (2003) proposed multiplying the conventional estimate of the variance of the coefficients given above by $\max(q, 1)$, where the *Knapp–Hartung variance modification factor* is

$$q = \frac{1}{n-k} \sum_i \frac{(y_i - \mathbf{x}_i \widehat{\boldsymbol{\beta}})^2}{\sigma_i^2 + \widehat{\tau}^2}$$

With the "empirical Bayes" estimator of $\widehat{\tau}^2$, $q = 1$, so this modification has no effect (Knapp and Hartung 2003).

7.4   Methods and formulas for predict

The *standard error of the prediction* (stdp) is $s_{p_i} = \sqrt{\mathbf{x}_i(\mathbf{X}'\widehat{\mathbf{V}}^{-1}\mathbf{X})^{-1}\mathbf{x}_i'}$.

The *leverages*, or diagonal elements of the projection matrix (hat), are

$$h_i = s_{p_i}^2/(\sigma_i^2 + \tau^2)$$

The *standard error of the forecast* (stdf) is $s_{f_i} = \sqrt{s_{p_i}^2 + \tau^2}$.

Denote the previously estimated coefficient vector by $\mathbf{b}$, and let $\lambda_i = \widehat{\tau}^2/(\sigma_i^2 + \widehat{\tau}^2)$ denote the empirical Bayes shrinkage factor for the ith observation.

The *predicted random effects* (u) are $u_i = \lambda_i(y_i - \mathbf{x}_i\mathbf{b})$.

The *standardized predicted random effects* (ustandard) are

$$u_{s_j} = (y_i - \mathbf{x}_i\mathbf{b})\left/\sqrt{\sigma_i^2 + \tau^2 - s_{p_i}^2}\right.$$

The *prediction including random effects* (xbu), or empirical Bayes estimate, is

$$\mathbf{x}_i\mathbf{b} + u_i = \lambda_i y_i + (1 - \lambda_i)\mathbf{x}_i\mathbf{b}$$

The *standard error of the prediction including random effects* (stdxbu) is

$$\sqrt{\lambda_i^2(\sigma_i^2 + \tau^2) + (1 - \lambda_i^2)s_{p_i}^2}$$

8   Acknowledgments

Stephen Sharp gave permission to release this package under the same name as his original Stata package for meta-regression and to incorporate his code. Debbie Lawlor gave permission to use the example dataset on exercise for depression and provided additional unpublished data. We thank Simon Thompson for his helpful comments on the manuscript, and we thank the organizers of and participants at a meeting in Park City, Utah, in 2005 for discussions that influenced the output displayed by metareg. Finally, we wish to thank the referee for helpful comments, which led to improvements in the program and the article.

9   References

Benjamini, Y., and Y. Hochberg. 1995. Controlling the false discovery rate: A practical and powerful approach to multiple testing. *Journal of the Royal Statistical Society, Series B (Methodological)* 57: 289–300.

Berkey, C. S., D. C. Hoaglin, F. Mosteller, and G. A. Colditz. 1995. A random-effects regression model for meta-analysis. *Statistics in Medicine* 14: 395–411.

Davey Smith, G., M. Egger, and A. N. Phillips. 1997. Meta-analysis: Beyond the grand mean? *British Medical Journal* 315: 1610–1614.

DerSimonian, R., and N. Laird. 1986. Meta-analysis in clinical trials. *Controlled Clinical Trials* 7: 177–188.

DuMouchel, W. H., and J. E. Harris. 1983. Bayes methods for combining the results of cancer studies in humans and other species. *Journal of the American Statistical Association* 78: 293–308.

Harris, R. J., M. J. Bradburn, J. J. Deeks, R. M. Harbord, D. G. Altman, and J. A. C. Sterne. 2008. metan: fixed- and random-effects meta-analysis. *Stata Journal* 8: 3–28. (Reprinted in this collection on pp. 29–54.)

Harville, D. A. 1977. Maximum likelihood approaches to variance component estimation and to related problems. *Journal of the American Statistical Association* 72: 320–338.

Higgins, J. P. T., and S. G. Thompson. 2004. Controlling the risk of spurious findings from meta-regression. *Statistics in Medicine* 23: 1663–1682.

Higgins, J. P. T., S. G. Thompson, J. J. Deeks, and D. G. Altman. 2002. Statistical heterogeneity in systematic reviews of clinical trials: A critical appraisal of guidelines and practice. *Journal of Health Services Research and Policy* 7: 51–61.

―――. 2003. Measuring inconsistency in meta-analyses. *British Medical Journal* 327: 557–560.

Higgins, J. P. T., S. G. Thompson, and D. J. Spiegelhalter. 2009. A re-evaluation of random-effects meta-analysis. *Journal of the Royal Statistical Society, Series A* 172: 137–159.

Knapp, G., and J. Hartung. 2003. Improved tests for a random-effects meta-regression with a single covariate. *Statistics in Medicine* 22: 2693–2710.

Lawlor, D. A., and S. W. Hopker. 2001. The effectiveness of exercise as an intervention in the management of depression: Systematic review and meta-regression analysis of randomised controlled trials. *British Medical Journal* 322: 763.

Lipsey, M. W., and D. B. Wilson. 2001. *Practical Meta-Analysis*. Thousand Oaks, CA: Sage.

Manly, B. F. J. 2006. *Randomization, Bootstrap and Monte Carlo Methods in Biology*. 3rd ed. Boca Raton, FL: Chapman & Hall/CRC.

Morris, C. N. 1983. Parametric empirical Bayes inference: Theory and applications. *Journal of the American Statistical Association* 78: 47–55.

Newson, R., and the ALSPAC Study Team. 2003. Multiple-test procedures and smile plots. *Stata Journal* 3: 109–132.

Shaffer, J. P. 1995. Multiple hypothesis testing. *Annual Review of Psychology* 46: 561–584.

Sharp, S. 1998. sbe23: Meta-analysis regression. *Stata Technical Bulletin* 42: 16–22. Reprinted in *Stata Technical Bulletin Reprints*, vol. 7, pp. 148–155. College Station, TX: Stata Press. (Reprinted in this collection on pp. 97–106.)

Sidik, K., and J. N. Jonkman. 2007. A comparison of heterogeneity variance estimators in combining results of studies. *Statistics in Medicine* 26: 1964–1981.

Sterne, J. A. C., and G. Davey Smith. 2001. Sifting the evidence—what's wrong with significance tests? *British Medical Journal* 322: 226–231.

Storey, J. D., J. E. Taylor, and D. Siegmund. 2004. Strong control, conservative point estimation and simultaneous conservative consistency of false discovery rates: a unified approach. *Journal of the Royal Statistical Society, Series B* 66: 187–205.

Sutton, A. J., and J. P. T. Higgins. 2008. Recent developments in meta-analysis. *Statistics in Medicine* 27: 625–650.

Thompson, S. G. 1993. Controversies in meta-analysis: The case of the trials of serum cholesterol reduction. *Statistical Methods in Medical Research* 2: 173–192.

Thompson, S. G., and J. P. T. Higgins. 2002. How should meta-regression analyses be undertaken and interpreted? *Statistics in Medicine* 21: 1559–1573.

———. 2005. Can meta-analysis help target interventions at individuals most likely to benefit? *Lancet* 365: 341–346.

Thompson, S. G., and S. J. Sharp. 1999. Explaining heterogeneity in meta-analysis: A comparison of methods. *Statistics in Medicine* 18: 2693–2708.

Wacholder, S., S. Chanock, M. Garcia-Closas, L. El ghormli, and N. Rothman. 2004. Assessing the probability that a positive report is false: an approach for molecular epidemiology studies. *Journal of the National Cancer Institute* 96: 434–442.

Westfall, P. H., and S. S. Young. 1993. *Resampling-Based Multiple Testing: Examples and Methods for p-Value Adjustment*. New York: Wiley.

Meta-analysis regression[1]

Stephen Sharp
MRC Epidemiology Unit
Institute of Metabolic Science
Cambridge, UK
stephen.sharp@mrc-epid.cam.ac.uk

The command `metareg` extends a random-effects meta-analysis to estimate the extent to which one or more covariates, with values defined for each study in the analysis, explain heterogeneity in the treatment effects. Such analysis is sometimes termed "meta-regression" (Lau, Ioannidis, and Schmid 1998). Examples of such study-level covariates might be average duration of follow-up, some measure of study quality, or, as described in this article, a measure of the geographical location of each study. `metareg` fits models with two additive components of variance, one representing the variance within units, the other the variance between units, and therefore is applicable both to the meta-analysis situation, where each unit is one study, and to other situations such as multicenter trials, where each unit is one center. Here `metareg` is explained in the meta-analysis context.

1 Background

Suppose y_i represents the treatment effect measured in study i (k independent studies, $i = 1, \ldots, k$), such as a log odds-ratio or a difference in means, v_i is the (within-study) variance of y_i, and $x_{i1}, \ldots, x_{ip}$ are measured study-level covariates. A weighted normal errors regression model is

$$Y \sim N(X\beta, V)$$

where $Y = (y_1, \ldots, y_k)^T$ is the $k \times 1$ vector of treatment effects, with ith element y_i, X is a $k \times (p+1)$ design matrix with ith row $(1, x_{i1}, \ldots, x_{ip})$, $\beta = (\beta_0, \ldots, \beta_p)^T$ is a $(p+1) \times 1$ vector of parameters, and V is a $k \times k$ diagonal variance matrix, with ith diagonal element v_i.

The parameters of this model can be estimated in Stata using `regress` with analytic weights $w_i = 1/v_i$. However, v_i represents the variance of the treatment effect within study i, so this model does not take into account any possible residual heterogeneity in the treatment effects *between* studies. One approach to incorporating residual heterogeneity is to include an additive between-study variance component τ^2, so the ith diagonal element of the variance matrix V becomes $v_i + \tau^2$.

The parameters of the model can then be estimated using a weighted regression with weights equal to $1/v_i + \tau^2$, but τ^2 must be explicitly estimated in order to carry out

---

1. This article uses the obsolete meta-analysis command `meta`, as well as obsolete graphics commands. The syntax described in the article still functions, although the updated syntax described in the article by Harbord and Higgins (2008) is preferred.—Ed.

the regression. `metareg` allows four alternative methods for estimation of τ^2, three of them are iterative, while one is noniterative and an extension of the moment estimator proposed for random effects meta-analysis without covariates (DerSimonian and Laird 1986).

2   Method-of-moments estimator

Maximum likelihood estimates of the β parameters are first obtained by weighted regression assuming $\widehat{\tau}^2 = 0$, and then a moment estimator of τ^2 is calculated using the residual sum of squares from the model,

$$\text{RSS} = \sum_{i=1}^{k} w_i(y_i - \widehat{y}_i)^2$$

as follows:

$$\widehat{\tau}^2_{\text{mm}} = \frac{\text{RSS} - \{k - (p+1)\}}{\sum_{i=1}^{k} w_i - \text{tr}\{V^{-1}X(X'V^{-1}X)^{-1}X'V^{-1}\}}$$

where $\widehat{\tau}^2_{\text{mm}} = 0$ if $\text{RSS} < k - (p+1)$ (DuMouchel and Harris 1983).

A weighted regression is then carried out with new weights $w_i^* = 1/\widehat{\tau}^2 + v_i$ to provide a new estimate of β. The formula for $\widehat{\tau}^2_{\text{mm}}$ in the case of no covariate reduces to the standard moment estimator (DerSimonian and Laird 1986).

3   Iterative procedures

Three other methods for estimating τ^2 have been proposed and require an iterative procedure.

Starting with $\widehat{\tau}^2 = 0$, a regression using weights $w_i^* = 1/v_i$ gives initial estimates of β. The fitted values $\widehat{y}_i$ from this model can then be used in one of three formulas for estimation of τ^2, given below:

$$\widehat{\tau}^2_{\text{ml}} = \frac{\sum_{i=1}^{k} w_i^{*2}\{(y_i - \widehat{y}_i)^2 - v_i\}}{\sum_{i=1}^{k} w_i^{*2}} \quad \text{maximum likelihood (Pocock, Cook, and Beresford}$$

1981)

$$\widehat{\tau}^2_{\text{reml}} = \frac{\sum_{i=1}^{k} w_i^{*2}\left\{\dfrac{k}{k - (p+1)}(y_i - \widehat{y}_i)^2 - v_i\right\}}{\sum_{i=1}^{k} w_i^{*2}} \quad \text{restricted maximum likelihood}$$

(Berkey et al. 1995)

$$\widehat{\tau}_{eb}^2 = \frac{\sum_{i=1}^{k} w_i^* \left\{ \dfrac{k}{k-(p+1)} (y_i - \widehat{y}_i)^2 - v_i \right\}}{\sum_{i=1}^{k} w_i^*} \quad \text{empirical Bayes (Berkey et al. 1995)}$$

In each case, if the estimated value $\widehat{\tau}^2$ is negative, it is set to zero.

Using the estimate $\widehat{\tau}^2$, new weights $w_i^* = 1/\widehat{\tau}^2 + v_i$ (or $1/v_i$ if $\widehat{\tau}$ is zero) are then calculated, and hence new estimates of β, fitted values $\widehat{y}_i$, and thence $\widehat{\tau}^2$. The procedure continues until the difference between successive estimates of τ^2 is less than a prespecified number (such as 10^{-4}). The standard errors of the final estimates of β are calculated forcing the scale parameter to be 1, since the weights are equal to the reciprocal variances.

4 Syntax

metareg has the usual syntax for a regression command, with the additional requirement that the user specify a variable containing either the within-study standard error or variance.

metareg *y varlist* $\left[\,if\,\right]$ $\left[\,in\,\right]$, {<u>wsse</u>(*varname*) | <u>wsv</u>ar(*varname*) |
      <u>wsse</u>(*varname*) <u>wsv</u>ar(*varname*)} $\left[\,\underline{bs}est(\{reml \mid ml \mid eb \mid mm\})\right.$
      <u>tol</u>eran(#) <u>lev</u>el(#) <u>noiter</u> $\left.\right]$

The command supplies estimated parameters, standard errors, Z statistics, p values and confidence intervals, in the usual regression output format. The estimated value of τ^2 is also given.

5 Options

wsse(*varname*) is a variable in the dataset, which contains the within-studies standard error $\sqrt{v_i}$. Either this or the wsvar option below (or both) must be specified.

wsvar(*varname*) is a variable in the dataset, which contains the within-studies variance v_i. Either this or the wsse option above (or both) must be specified.

Note: If both the above options are specified, the program will check that the variance is the square of the standard error for each study.

bsest({reml | ml | eb | mm}) specifies the method for estimating τ^2. The default is reml (restricted maximum likelihood), with the alternatives being ml (maximum likelihood), eb (empirical Bayes), and mm (method of moments).

toleran(#) specifies the difference between values of $\widehat{\tau}^2$ at successive iterations required for convergence. If # is n, the process will not converge until successive values of $\widehat{\tau}^2$ differ by less than 10^{-n}. The default is 4.

level(*#*) specifies the confidence level, as a percentage, for confidence intervals. The default is level(95) or as set by set level; see [U] **20.7 Specifying the width of confidence intervals**.

noiter requests that the log of the iterations in the reml, ml, or eb procedures be suppressed from the output.

6   Example

BCG is a vaccine widely used to give protection against tuberculosis. Colditz et al. (1994) performed a meta-analysis of all published trials, which randomized subjects to either BCG vaccine or placebo, and then had similar surveillance procedures to monitor the outcome, diagnosis of tuberculosis.

The data in bcg.dta are as reported by Berkey et al. (1995). Having read the file into Stata, the log odds-ratio of tuberculosis comparing BCG with placebo, and its standard error can be calculated for each study.

```
. use bcg, clear
(BCG and tuberculosis)

. describe

Contains data from bcg.dta
  obs:            13                          BCG and tuberculosis
 vars:             8
 size:           351
-------------------------------------------------------------------------------
    1. trial       str2    %9s          trial identity number
    2. lat         byte    %9.0g        absolute latitude from Equator
    3. nt          float   %9.0g        total vaccinated patients
    4. nc          float   %9.0g        total unvaccinated patients
    5. rt          int     %9.0g        tuberculosis in vaccinated
    6. rc          int     %9.0g        tuberculosis in unvaccinated
-------------------------------------------------------------------------------
Sorted by:

. list, noobs
       trial      lat        nt        nc       rt        rc
           1       44       123       139        4        11
           2       55       306       303        6        29
           3       42       231       220        3        11
           4       52     13598     12867       62       248
           5       13      5069      5808       33        47
           6       44      1541      1451      180       372
           7       19      2545       629        8        10
           8       13     88391     88391      505       499
           9       27      7499      7277       29        45
          10       42      1716      1665       17        65
          11       18     50634     27338      186       141
          12       33      2498      2341        5         3
          13       33     16913     17854       27        29
. generate logor=log((rt/(nt-rt))/(rc/(nc-rc)))
. generate selogor=sqrt((1/rc)+(1/(nc-rc))+(1/rt)+(1/(nt-rt)))
```

Note: If either `rt` or `rc` were 0, a standard approach would be to add 0.5 to each of `rt`, `rc`, `nt-rt`, and `nc-rc` for that study (Cox and Snell 1989).

A meta-analysis of the data can now be performed using the `meta` command described by Sharp and Sterne (1997, 1998).

```
. meta logor selogor, eform graph(r) id(trial) cline xlab(0.5,1,1.5) xline(1)
> boxsh(4) b2("Odds ratio - log scale")
Meta-analysis (exponential form)
          | Pooled      95% CI        Asymptotic      No. of
 Method   |   Est  Lower   Upper   z_value  p_value   studies
 ---------+-----------------------------------------------------
 Fixed    | 0.647  0.595   0.702   -10.319   0.000      13
 Random   | 0.474  0.325   0.690   -3.887    0.000
Test for heterogeneity: Q= 163.165 on 12 degrees of freedom (p= 0.000)
Moment estimate of between-studies variance =  0.366
```

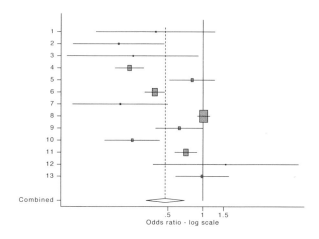

Figure 1. A meta-analysis of the BCG and Tuberculosis data

Both the graph and the statistical test indicate substantial heterogeneity between the trials, with an estimated between-studies variance of 0.366. The random effects combined estimate of 0.474, indicating a strong protective effect of BCG against tuberculosis, should not be reported without some discussion of the possible reasons for the differences between the studies (Thompson 1994).

One possible explanation for the differences in treatment effects could be that the studies were conducted at different latitudes from the equator. Berkey et al. (1995) speculated that absolute latitude, or distance of each study from the equator, may serve as a surrogate for the presence of environmental mycobacteria, which provide a certain level of natural immunity against tuberculosis. By sorting on absolute latitude, the graph obtained using `meta` shows the studies in order of increasing latitude going down the page.

```
. sort lat
. meta logor selogor, eform graph(r) id(trial) cline xlab(0.5,1,1.5) xline(1)
> boxsh(4) b2("Odds ratio - log scale")
```

(output omitted)

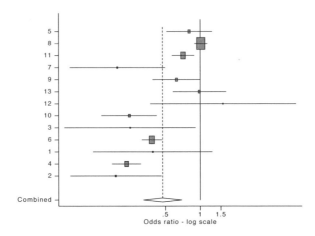

Figure 2. Same as figure 1 but sorted by latitude

The graph now suggests that BCG vaccination is more effective at higher absolute latitudes. This can be investigated further using the `metareg` command, with a REML estimate of the between-studies variance τ^2.

```
. metareg logor lat, wsse(selogor) bs(reml) noiter
Meta-analysis regression                          No of studies =    13
                                                  tau^2 method        reml
                                                  tau^2 estimate =  .0235

Successive values of tau^2 differ by less than 10^-4 - convergence achieved
```

	Coef.	Std. Err.	z	P>|z|	[95% Conf. Interval]
lat	-.0320363	.0049432	-6.481	0.000	-.0417247 -.0223479
_cons	.3282194	.1659807	1.977	0.048	.0029033 .6535356

This analysis shows that after allowing for additive residual heterogeneity, there is a significant negative association between the log odds-ratio and absolute latitude, i.e., the higher the absolute latitude, the lower the odds ratio, and hence the greater the benefit of BCG vaccination. The following plot of log odds-ratio against absolute latitude includes the fitted regression line from the model above. The size of the circles in the plot is inversely proportional to the variance of the log odds-ratio, so larger circles correspond to larger studies.

```
. generate invvlor=selogor^-2

. generate fit=0.328-0.032*lat

. graph logor fit lat [fw=invvlor], s(oi) c(.1)  xlab(0,10,20,30,40,50,60)
> ylab(-1.6094,-0.6931,0,0.6931) ll("Odds ratio (log scale)")
> b2("Distance from Equator (degrees of latitude)")
```

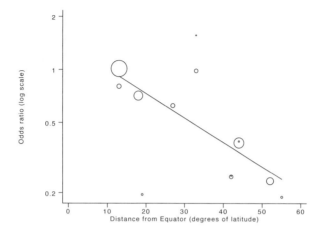

Figure 3. Plot of log odds-ratio against absolute latitude

(Note: The axes on this graph have been modified using the STAGE software)

Here a restricted maximum-likelihood method was used to estimate τ^2; the other three methods are used in turn below:

```
. metareg logor lat, wsse(selogor) bs(ml) noiter
Meta-analysis regression                         No of studies =   13
                                                 tau^2 method       ml
                                                 tau^2 estimate =  .0037
Successive values of tau^2 differ by less than 10^-4 - convergence achieved
-----------------------------------------------------------------------------
          |    Coef.   Std. Err.       z     P>|z|     [95% Conf. Interval]
----------+------------------------------------------------------------------
      lat | -.0327447   .0033327     -9.825   0.000    -.0392767   -.0262128
    _cons |  .3725098   .1043895      3.568   0.000     .1679102    .5771093
-----------------------------------------------------------------------------
. metareg logor lat, wsse(selogor) bs(eb) noiter
Meta-analysis regression                         No of studies =   13
                                                 tau^2 method       eb
                                                 tau^2 estimate =  .1373
Successive values of tau^2 differ by less than 10^-4 - convergence achieved
-----------------------------------------------------------------------------
          |    Coef.   Std. Err.       z     P>|z|     [95% Conf. Interval]
----------+------------------------------------------------------------------
      lat | -.0305794   .0090005     -3.398   0.001    -.0482201   -.0129388
    _cons |  .2548214   .3138935      0.812   0.417    -.3603984    .8700413
-----------------------------------------------------------------------------
```

```
. metareg logor lat, bs(mm) wsse(selogor) noiter
Warning: mm is a non-iterative method, noiter option ignored
Meta-analysis regression                                No of studies =    13
                                                        tau^2 method       mm
                                                        tau^2 estimate =  .0480

-------------------------------------------------------------------------------
           |     Coef.    Std. Err.       z     P>|z|     [95% Conf. Interval]
---------+---------------------------------------------------------------------
       lat | -.0315724    .0061726    -5.115   0.000     -.0436704   -.0194744
     _cons |   .303035    .2108751     1.437   0.151     -.1102727    .7163427
-------------------------------------------------------------------------------
```

The estimated value of τ^2 using a method-of-moments estimator is 0.048, compared with 0.366 before adjusting for latitude, so absolute latitude has explained almost all of the variation between the studies.

The analyses above show that the estimate of the effect of latitude is similar using all four methods. However, the estimated values of τ^2 differ considerably, with the estimate from the empirical Bayes method being largest. The restricted maximum-likelihood method corrects the bias in the maximum-likelihood estimate of τ^2. The basis for using the empirical Bayes method is less clear (Morris 1983), so this method should be used with caution. The moment-based method extends the usual random-effects meta-analysis; below metareg is used to fit a model with no covariate:

```
. metareg logor, bs(mm) wsse(selogor)
Meta-analysis regression                                No of studies =    13
                                                        tau^2 method       mm
                                                        tau^2 estimate =  .3663

-------------------------------------------------------------------------------
           |     Coef.    Std. Err.       z     P>|z|     [95% Conf. Interval]
---------+---------------------------------------------------------------------
     _cons | -.7473923    .1922628    -3.887   0.000     -1.124221   -.3705641
-------------------------------------------------------------------------------
```

Now the estimate of τ^2 is identical to that obtained earlier from meta, and the constant parameter is the log of the random effects pooled estimate given by meta.

The paper by Thompson and Sharp (1999) contains a fuller discussion both of the differences between the four methods of estimation, and other methods for explaining heterogeneity. Copies are available on request from the author.

7   Saved results

metareg saves the following results in the S_ macros:

S_1 k, number of studies
S_2 $\hat\tau^2$, estimate of between-studies variance

8 Acknowledgments

I am grateful to Simon Thompson, Ian White, and Jonathan Sterne for their helpful comments on earlier versions of this command.

9 References

Berkey, C. S., D. C. Hoaglin, F. Mosteller, and G. A. Colditz. 1995. A random-effects regression model for meta-analysis. *Statistics in Medicine* 14: 395–411.

Colditz, G. A., T. F. Brewer, C. S. Berkey, M. E. Wilson, E. Burdick, H. V. Fineberg, and F. Mosteller. 1994. Efficacy of BCG vaccine in the prevention of tuberculosis. Meta-analysis of the published literature. *Journal of the American Medical Association* 271: 698–702.

Cox, D. R., and E. J. Snell. 1989. *Analysis of Binary Data.* 2nd ed. London: Chapman & Hall.

DerSimonian, R., and N. Laird. 1986. Meta-analysis in clinical trials. *Controlled Clinical Trials* 7: 177–188.

DuMouchel, W. H., and J. E. Harris. 1983. Bayes methods for combining the results of cancer studies in humans and other species. *Journal of the American Statistical Association* 78: 293–308.

Harbord, R. M., and J. P. T. Higgins. 2008. Meta-regression in Stata. *Stata Journal* 8: 493–519. (Reprinted in this collection on pp. 70–96.)

Lau, J., J. P. A. Ioannidis, and C. H. Schmid. 1998. Summing up evidence: One answer is not always enough. *Lancet* 351: 123–127.

Morris, C. N. 1983. Parametric empirical Bayes inference: Theory and applications. *Journal of the American Statistical Association* 78: 47–55.

Pocock, S. J., D. G. Cook, and S. A. A. Beresford. 1981. Regression of area mortality rates on explanatory variables: What weighting is appropriate? *Applied Statistics* 30: 286–295.

Sharp, S., and J. A. C. Sterne. 1997. sbe16: Meta-analysis. *Stata Technical Bulletin* 38: 9–14. Reprinted in *Stata Technical Bulletin Reprints*, vol. 7, pp. 100–106. College Station, TX: Stata Press.[1]

———. 1998. sbe16.1: New syntax and output for the meta-analysis command. *Stata Technical Bulletin* 42: 6–8. Reprinted in *Stata Technical Bulletin Reprints*, vol. 7, pp. 106–108. College Station, TX: Stata Press.[1]

---

1. The original command to perform meta-analysis was meta, documented in the sbe16 articles; meta is now metan. metan is described in an updated article, sbe24, on pages 3–28 of this collection.—Ed.

Thompson, S. G. 1994. Why sources of heterogeneity in meta-analysis should be investigated. *British Medical Journal* 309: 1351–1355.

Thompson, S. G., and S. J. Sharp. 1999. Explaining heterogeneity in meta-analysis: A comparison of methods. *Statistics in Medicine* 18: 2693–2708.

Part 3

Investigating bias in meta-analysis: metafunnel, confunnel, metabias, and metatrim

A series of empirical studies of the medical literature have established that flaws in the conduct of randomized trials lead to exaggeration of intervention effect estimates and that publication and other reporting biases lead to statistically significant results being more likely to be reported. These biases have the potential to cause the results of meta-analyses to be overoptimistic. There is longstanding interest in graphical and statistical methods to detect publication and other biases, but the application and interpretation of the methods is controversial. Therefore, users are recommended to consult published guidance; for example, see Sterne, Egger, and Moher (2008).

Funnel plots, which can be displayed using the `metafunnel` command, have long been recommended as a graphical display that can be used to examine whether the results of a meta-analysis are affected by publication bias, because publication bias can lead to an association between study size and the size of effect, and hence to an asymmetric appearance of a funnel plot. However, there are a number of causes of such small-study effects (Egger et al. 1997; Sterne, Gavaghan, and Egger 2000), and these should be carefully considered when funnel plots appear asymmetric. Contour-enhanced funnel plots can be displayed using the `confunnel` command by Palmer et al. These aim to aid interpretation by shading areas of the funnel plot corresponding to different areas of statistical significance.

Visual assessment of funnel plots is inherently subjective, and so a number of authors have proposed statistical tests for funnel plot asymmetry that are available in the `metabias` command. The appropriate use of such tests has been much discussed by statisticians, because the most widely used test (Egger et al. 1997) can give false-positive results in some circumstances. These issues are discussed in the article by Harbord et al. describing the most recent version of `metabias`, which implements new tests for funnel plot asymmetry that aim to overcome some of the problems with the test proposed by Egger et al. (1997). The `metabias` command was originally written by Steichen, and his articles describing the tests implemented in the original versions of the command are also included in this collection. Steichen also wrote the `metatrim` command, which implements the "trim and fill" approach to identifying and correcting for funnel plot asymmetry arising from publication bias (Taylor and Tweedie 1998; Duval and Tweedie 2000). This approach has also been much debated—there are suggestions that it performs poorly in the presence of substantial between-study heterogeneity (Terrin et al. 2003; Peters et al. 2007).

1   References

Duval, S., and R. Tweedie. 2000. Trim and fill: A simple funnel-plot–based method of testing and adjusting for publication bias in meta-analysis. *Biometrics* 56: 455–463.

Egger, M., G. Davey Smith, M. Schneider, and C. Minder. 1997. Bias in meta-analysis detected by a simple, graphical test. *British Medical Journal* 315: 629–634.

Peters, J. L., A. J. Sutton, D. R. Jones, K. R. Abrams, and L. Rushton. 2007. Performance of the trim and fill method in the presence of publication bias and between-study heterogeneity. *Statistics in Medicine* 26: 4544–4562.

Sterne, J. A. C., M. Egger, and D. Moher. 2008. Addressing reporting biases. In *Cochrane Handbook for Systematic Reviews of Interventions*, ed. J. P. T. Higgins and S. Green, 297–334. Chichester, UK: Wiley.

Sterne, J. A. C., D. Gavaghan, and M. Egger. 2000. Publication and related bias in meta-analysis: Power of statistical tests and prevalence in the literature. *Journal of Clinical Epidemiology* 53: 1119–1129.

Taylor, S. J., and R. L. Tweedie. 1998. Practical estimates of the effect of publication bias in meta-analysis. *Australasian Epidemiologist* 5: 14–17.

Terrin, N., C. H. Schmid, J. Lau, and I. Olkin. 2003. Adjusting for publication bias in the presence of heterogeneity. *Statistics in Medicine* 22: 2113–2126.

The Stata Journal (2004)
4, Number 2, pp. 127–141

Funnel plots in meta-analysis

Jonathan A. C. Sterne and Roger M. Harbord
Department of Social Medicine
University of Bristol
Bristol, UK
roger.harbord@bristol.ac.uk

Abstract. Funnel plots are a visual tool for investigating publication and other bias in meta-analysis. They are simple scatterplots of the treatment effects estimated from individual studies (horizontal axis) against a measure of study size (vertical axis). The name "funnel plot" is based on the precision in the estimation of the underlying treatment effect increasing as the sample size of component studies increases. Therefore, in the absence of bias, results from small studies will scatter widely at the bottom of the graph, with the spread narrowing among larger studies. Publication bias (the association of publication probability with the statistical significance of study results) may lead to asymmetrical funnel plots. It is, however, important to realize that publication bias is only one of a number of possible causes of funnel plot asymmetry—funnel plots should be seen as a generic means of examining small study effects (the tendency for the smaller studies in a meta-analysis to show larger treatment effects) rather than a tool to diagnose specific types of bias. This article introduces the `metafunnel` command, which produces funnel plots in Stata. In accordance with published recommendations, standard error is used as the measure of study size. Treatment effects expressed as ratio measures (for example risk ratios or odds ratios) may be plotted on a log scale.

Keywords: st0061, metafunnel, funnel plots, meta-analysis, publication bias, small-study effects

1 Introduction

The substantial recent interest in meta-analysis (the statistical methods that are used to combine results from a number of different studies) is reflected in a number of user-written commands that do meta-analysis in Stata. Meta-analyses should be based on *systematic reviews* of relevant literature. A systematic review is a systematic assembly, critical appraisal, and synthesis of all relevant studies on a specific topic. The main feature that distinguishes systematic from narrative reviews is a methods section that clearly states the question to be addressed and the methods and criteria to be employed for identifying and selecting relevant studies and extracting and analyzing information (Egger, Davey Smith, and Altman 2001).

While systematic reviews and meta-analyses have the potential to produce precise estimates of treatment effects that reflect all of the relevant literature, they are not immune to bias. *Publication bias*—the association of publication probability with the statistical significance of study results—is well documented as a problem in the medical research literature (Stern and Simes 1997). Further, it has been demonstrated that randomized controlled trials for which concealment of treatment allocation is not adequate, or which are not double blind, produce estimated treatment effects that appear more beneficial (Schulz et al. 1995).

2   Funnel plots

Funnel plots are simple scatterplots of the treatment effects estimated from individual studies against a measure of study size. The name "funnel plot" is based on the precision in the estimation of the underlying treatment effect increasing as the sample size of component studies increases. Results from small studies will therefore scatter widely at the bottom of the graph, with the spread narrowing among larger studies. In the absence of bias, the plot will resemble a symmetrical, inverted funnel, as shown in the top graph of figure 1.

If there is bias, for example, because smaller studies showing no statistically significant effects (open circles in figure 1) remain unpublished, then such publication bias will lead to an asymmetrical appearance of the funnel plot with a gap in the right bottom side of the graph (middle graph of figure 1). In this situation, the combined effect from meta-analysis will overestimate the treatment's effect. The more pronounced the asymmetry, the more likely it is that the amount of bias will be substantial.

It is important to realize that publication bias is only one of a number of possible explanations for funnel plot asymmetry; these are discussed in more detail in section 2.3. For example, trials of lower quality yield exaggerated estimates of treatment effects (Schulz et al. 1995). Smaller studies are, on average, conducted and analyzed with less methodological rigor than larger studies (Egger et al. 2003), so asymmetry may also result from the overestimation of treatment effects in smaller studies of lower methodological quality (bottom graph of figure 1). Unfortunately, funnel plot asymmetry has often been equated with publication bias without consideration of its other possible explanations; for example, the help file for the `metabias` command in Stata (written in 1998) refers only to publication bias.

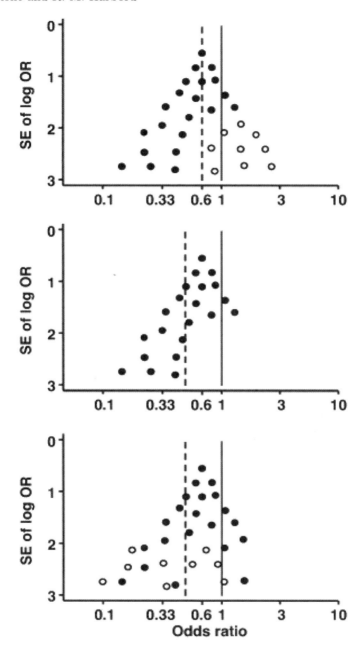

Figure 1. Hypothetical funnel plots: (top) symmetrical plot in the absence of bias (open circles indicate smaller studies showing no beneficial effects); (middle) asymmetrical plot in the presence of publication bias (smaller studies showing no beneficial effects are missing); (bottom) asymmetrical plot in the presence of bias due to low methodological quality of smaller studies (open circles indicate small studies of inadequate quality whose results are biased towards larger beneficial effects).

Although it is conventional to plot treatment effects on the horizontal axis and the measure of study size on the vertical axis, it is certainly not an error to plot the axes the other way around. Indeed, such a choice is arguably more consistent with standard statistical practice in that the variable on the vertical axis is usually hypothesized to depend on the variable on the horizontal axis. Such funnel plots can be plotted in Stata using the `metabias` command (Steichen 1998; Steichen, Egger, and Sterne 1998).

2.1   Choice of axis in funnel plots

The majority of endpoints in randomized trials of medical treatments are binary, with treatment effects most commonly expressed as ratio measures (odds ratio, risk ratio, or hazard ratio). (This may not be true of trials in other disciplines, such as psychology or social research.) The use of ratio measures is justified by empirical evidence that there is less between-trial heterogeneity in treatment effects based on ratio measures than difference measures (Deeks and Altman 2001; Engels et al. 2000). As is generally the case in meta-analysis, the *log* of the ratio measure and its standard error are used in funnel plots.

Sterne and Egger (2001) consider choice of axis in funnel plots of meta-analyses with binary outcomes. Although sample size or functions of sample size have often been used as the vertical axis, this is problematic because the precision of a treatment effect estimate is determined by both the sample size and by the number of events. Thus, studies with very different sample sizes may have the same standard error and precision and vice versa. Therefore, the shape of plots using sample size on the vertical axis is not predictable except that, in the absence of bias, it should be symmetric. After considering various possible choices of vertical axis, Sterne and Egger conclude that standard error of the treatment effect estimate is likely to be preferable in many situations. Funnel plots may also be drawn using precision (= 1/(standard error)) on the vertical axis using the `funnel2` command distributed as part of the `metaggr` package (Bradburn, Deeks, and Altman 1998). Such plots tend to emphasize differences between the largest study and the others.

2.2   Example

The trials of magnesium therapy following myocardial infarction (heart attack) are a well-known example in which the results of a meta-analysis, which appeared to provide clear evidence that magnesium therapy reduced mortality, were contradicted by subsequent larger trials that found no evidence that magnesium influenced mortality. Figure 2 is a funnel plot based on the results of 15 trials of the effect of magnesium on mortality following myocardial infarction. Because the smaller trials produced smaller odds ratios (more substantial reductions in mortality associated with magnesium therapy), the funnel plot is clearly asymmetric.

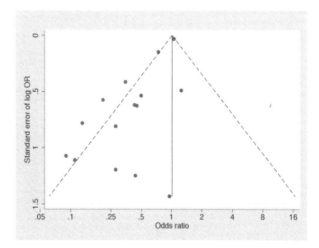

Figure 2. Funnel plot, using data from 15 trials of magnesium therapy following myocardial infarction.

The horizontal axis of figure 2 (treatment odds ratio) is drawn on a log scale, so that (for example) odds ratios of 2 and 0.5 are the same distance from the null value of 1 (no treatment effect). This is equivalent to plotting the log odds-ratio on the horizontal axis. The standard error of the log OR is plotted on the vertical axis. Note that the largest studies have the smallest standard errors, so to place the largest studies at the top of the graph, the vertical axis must be reversed (standard error 0 at the top).

The solid vertical line represents the summary estimate of the treatment effect, derived using fixed-effects meta-analysis. This is close to 1 because the estimated treatment odds ratios in the largest studies were close to 1. For the purposes of displaying the center of the plot in the absence of bias, calculation of the summary log odds-ratio using fixed rather than random-effects meta-analysis is preferable because the random-effects estimate gives greater relative weight to smaller studies and will, therefore, be more affected if publication bias is present (Poole and Greenland 1999).

Interpretation of funnel plots is facilitated by inclusion of diagonal lines representing the 95% confidence limits around the summary treatment effect, i.e., [summary effect estimate − (1.96 × standard error)] and [summary effect estimate + (1.96 × standard error)] for each standard error on the vertical axis. These show the expected distribution of studies in the absence of heterogeneity or of selection biases: in the absence of heterogeneity, 95% of the studies should lie within the funnel defined by these straight lines. Because these lines are not strict 95% limits, they are referred to as "pseudo 95% confidence limits".

2.3 Sources of funnel plot asymmetry

Funnel plots were first proposed as a means of detecting a specific form of bias— publication bias. However as explained earlier (see the bottom graph of figure 1), the exaggeration of treatment effects in small studies of low quality provides a plausible alternative mechanism for funnel plot asymmetry. Egger et al. (1997) list different possible reasons for funnel plot asymmetry, which are summarized in table 1.

Table 1. Potential sources of asymmetry in funnel plots

1. Selection biases
    Publication bias
        Location biases
            Language bias
            Citation bias
            Multiple publication bias

2. True heterogeneity
    Size of effect differs according to study size:
        Intensity of intervention
        Differences in underlying risk

3. Data irregularities
    Poor methodological design of small studies
    Inadequate analysis
    Fraud

4. Artifact
    Heterogeneity due to poor choice of effect measure

5. Chance

In addition to selective publication of studies according to their results, other possible biases affecting the selection of studies for inclusion in meta-analyses include the propensity for the results to affect the language of publication (Jüni et al. 2002); the possibility that results affect the frequency with which a study is cited and, hence, its probability of inclusion in a meta-analysis, and the multiple publication of studies with demonstrating an effect of the intervention (Tramer et al. 1997).

It is important to realize that funnel plot asymmetry need not result from bias. The studies displayed in a funnel plot may not always estimate the same underlying effect of the same intervention, and such heterogeneity in results may lead to asymmetry in funnel plots if the true treatment effect is larger in the smaller studies. For example, if a combined outcome is considered, then substantial benefit may be seen only in subjects at high risk for the component of the combined outcome which is affected by the intervention (Davey Smith and Egger 1994; Glasziou and Irwig 1995). Some interventions may have been implemented less thoroughly in larger studies, thus explaining the more positive results in smaller studies. For example, an asymmetrical funnel plot was found in a meta-analysis of trials examining the effect of inpatient comprehensive geriatric assessment programs on mortality. An experienced consultant geriatrician was more likely to be actively involved in the smaller trials and this may explain the larger treatment effects observed in these trials (Egger et al. 1997; Stuck et al. 1993).

The way in which data irregularities such as low methodological quality of smaller studies may result in funnel plot asymmetry was described earlier. Poor choice of effect measure may also result in funnel plot asymmetry; for example, it has been shown that meta-analyses in which intervention effects are measured as risk differences are more heterogeneous than those in which intervention effects are measured as risk ratios or odds ratios (Deeks and Altman 2001; Engels et al. 2000). The inappropriate use of risk differences may also result in funnel plot asymmetry—if the effect of intervention is homogeneous on the risk-ratio scale, then the risk difference will be smaller in studies that have low event rates.

2.4   Tests for funnel plot asymmetry

It is, of course, possible that an asymmetrical funnel plot arises merely by the play of chance. Statistical tests for funnel plot asymmetry have been proposed by Begg and Mazumdar (1994) and by Egger et al. (1997). These are available in the Stata command `metabias` (Steichen 1998; Steichen, Egger, and Sterne 1998). The test proposed by Egger et al. (1997) is algebraically identical to a test that there is no linear association between the treatment effect and its standard error and, hence, that there is no straight-line association in the funnel plot of treatment effect against its standard error (see Sterne, Gavaghan, and Egger [2000] for details). The corresponding fitted line may be added to the funnel plot using the `egger` option of the `metafunnel` command—see section 5 below.

2.5   Small-study effects

Funnel plot asymmetry thus raises the possibility of bias, but it is not proof of bias. It is important to note, however, that asymmetry (unless produced by chance alone) will always lead us to question the interpretation of the overall estimate of effect when studies are combined in a meta-analysis; for example, if the study size predicts the treatment effect, what treatment effect will apply if the treatment is adopted in routine practice? Sterne, Egger, and Davey Smith (2001) and Sterne, Gavaghan, and Egger (2000) have suggested that the funnel plot should be seen as a generic means of examining "small-study effects" (the tendency for the smaller studies in a meta-analysis to show larger treatment effects) rather than as a tool to diagnose specific types of bias.

When funnel plot asymmetry is found, its possible causes should be carefully considered. For example, how comprehensive was the literature search that located the trials included in the meta-analysis? Does reported trial quality differ between larger and smaller studies? Is there a plausible reason for the effect of intervention to be greater in smaller trials? It is possible that differences between smaller and larger trials are accounted for by a trial characteristic; this may be investigated using the by() option of the metafunnel command, as described in section 6 below. Explanations for heterogeneity may be investigated more formally using meta-regression (Thompson and Sharp 1999) to investigate associations between study characteristics and intervention effect estimates. For example, we might investigate evidence that studies in which reported allocation concealment is unclear or inadequate tend to result in more beneficial treatment effect estimates. Meta-regression analyses may be done using the Stata command metareg (Sharp 1998); however, it will not necessarily be possible to provide a definitive explanation for funnel plot asymmetry. In medical research, meta-analyses typically contain 10 or fewer trials (Sterne, Gavaghan, and Egger 2000). Power to detect associations between study characteristics and intervention effect estimates will therefore often be low, in which case it may not be possible to identify a particular study characteristic as the cause of the heterogeneity.

3   Syntax

metafunnel { *theta* { *se* | *var* } | *exp(theta)* { *ll ul* [*cl*] } } [*if*] [*in*]

    [, by(*by_var*) [var | ci] nolines forcenull reverse eform egger

    graph_options]

4   Description

metafunnel plots funnel plots. The syntax for metafunnel is based on the same framework as for the meta, metabias, metacum, and metatrim commands. The user provides the effect estimate as *theta* (e.g., the log odds-ratio) and a measure of theta's variability (i.e., its standard error or its variance). Alternatively, the user provides *exp(theta)* (e.g., an odds ratio), its confidence interval, and, optionally, the confidence level.

5   Options

by(*by_var*) displays subgroups according to the value of *by_var*. The legend displays the value labels for the levels of *by_var* if these are present; otherwise, it displays the value of each level of *by_var*.

var and ci indicate the meaning of the input variables in the same way as for the other meta-analysis commands listed above. The help file for meta gives a full explanation.

nolines specifies that pseudo 95% confidence interval lines not be included in the plot. The default is to include them.

forcenull forces the vertical line at the center of the funnel to be plotted at the null treatment effect of zero (1 when the treatment effect is exponentiated). The default is for the line to be plotted at the value of the fixed-effects summary estimate.

reverse inverts the funnel plot so that larger studies are displayed at the bottom of the plot with smaller studies at the top. This may also be achieved by specifying noreverse as part of the yscale(*axis_description*) graphics option.

eform exponentiates the treatment effect *theta* and displays the horizontal axis (treatment effect) on a log scale. As discussed in section 2.2, this is useful for displaying ratio measures, such as odds ratios and risk ratios.

egger adds the fitted line corresponding to the regression test for funnel plot asymmetry proposed by Egger et al. (1997) and implemented in metabias (see section 2.4). This option may not be combined with the by() option.

graph_options are any options allowed by the twoway scatter command that can be used to change the appearance of the points and add labels. If option egger is specified, the look of the fitted line can be changed using the options clstyle, clpattern, clwidth, and clcolor explained under *connect_options* in Stata's built-in help system and the graphics manual.

6 Examples

Listing the data for the 15 magnesium trials produces the following output:

```
. list trial trialnam year dead1 alive1 dead0 alive0, noobs
```

trial	trialnam	year	dead1	alive1	dead0	alive0
1	Morton	1984	1	39	2	34
2	Rasmussen	1986	9	126	23	112
3	Smith	1986	2	198	7	193
4	Abraham	1987	1	47	1	45
5	Feldstedt	1988	10	140	8	140
6	Schechter	1989	1	58	9	47
7	Ceremuzynski	1989	1	24	3	20
8	Singh	1990	6	70	11	64
9	Pereira	1990	1	26	7	20
10	Schechter 1	1991	2	87	12	68
11	Golf	1991	5	18	13	20
12	Thogersen	1991	4	126	8	114
13	LIMIT-2	1992	90	1069	118	1039
14	Schechter 2	1995	4	103	17	91
15	ISIS-4	1995	2216	26795	2103	26936

To use the metafunnel command, we first need to derive the treatment effect and its standard error for each trial. Here, we will express the treatment effects as log odds-ratios.

```
. generate or = (dead1/alive1)/(dead0/alive0)
. generate logor = log(or)
. generate selogor = sqrt((1/dead1)+(1/alive1)+(1/dead0)+(1/alive0))
```

A funnel plot can then be drawn using the following syntax, which includes the regression line corresponding to the regression test for funnel plot asymmetry proposed by Egger et al. (1997):

```
. metafunnel logor selogor, xtitle(Log odds ratio) ytitle(Standard error of log OR)
> egger
```

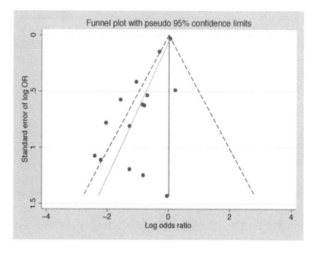

Figure 3. Funnel plot, using data from 15 trials of magnesium therapy following my-ocardial infarction, with log odds-ratios displayed on the horizontal axis.

By default, the subtitle "Funnel plot with pseudo 95% confidence limits" is displayed. ("Funnel plot" is displayed if the **nolines** options is specified.) This may be changed using the graphics option **subtitle**(*tinfo*).

Note that the log odds-ratio and its standard error may be derived automatically using the **metan** command. (The latest version of this command may be installed by typing **ssc install metaaggr.pkg, replace** in the Stata Command window.) Typing

```
. metan dead1 alive1 dead0 alive0, or
```

produces a meta-analysis of the effect of magnesium and creates variables _ES, containing the odds ratio in each study, and _selogES, containing the standard error of the log odds-ratio. Thus, we may derive the log odds-ratio by typing

```
. generate log_ES = log(_ES)
```

The list output below shows that variables log_ES _selogES are identical to variables **logor** and **selogor** derived earlier.

```
. list trial trialnam year logor selogor _ES log_ES _selogES, noobs
```

trial	trialnam	year	logor	selogor	_ES	log_ES	_selogES
1	Morton	1984	-.8303483	1.247018	.4358974	-.8303483	1.247018
2	Rasmussen	1986	-1.056053	.4140706	.3478261	-1.056053	.4140706
3	Smith	1986	-1.27834	.8081392	.2784993	-1.27834	.8081392
4	Abraham	1987	-.0434851	1.42951	.9574468	-.0434851	1.42951
5	Feldstedt	1988	.2231435	.4891684	1.25	.2231435	.4891684
6	Schechter	1989	-2.40752	1.072208	.0900383	-2.40752	1.072208
7	Ceremuzynski	1989	-1.280934	1.193734	.2777778	-1.280934	1.193734
8	Singh	1990	-.695748	.5361776	.4987013	-.695748	.5361776
9	Pereira	1990	-2.208274	1.109648	.1098901	-2.208274	1.109648
10	Schechter 1	1991	-2.03816	.7807263	.1302682	-2.03816	.7807263
11	Golf	1991	-.8501509	.6184486	.4273504	-.8501509	.6184486
12	Thogersen	1991	-.7932307	.6258662	.452381	-.7932307	.6258662
13	LIMIT-2	1992	-.2993398	.1465729	.7413074	-.2993398	.1465729
14	Schechter 2	1995	-1.570789	.5740395	.2078812	-1.570789	.5740395
15	ISIS-4	1995	.0575872	.0316421	1.059278	.0575872	.0316421

The following command was used to produce figure 2 (see section 2.2), in which the horizontal axis is the treatment odds ratio, displayed on a log scale:

```
. metafunnel logor selogor, xlab(.05 .1 .25 .5 1 2 4 8 16)
> xscale(log) xtitle(Odds ratio) eform subtitle( )
> ytitle(Standard error of log OR)
```

When the `eform` option is used, the label of the horizontal axis (treatment effect, *theta*) is changed accordingly, unless there is a variable label for *theta* or the `xtitle(`*axis_title*`)` graphics option is used.

Finally, we will illustrate the use of the `by()` option by grouping the studies according to whether they were published during the 1980s or the 1990s:

```
. generate period = year
. recode period 1980/1989=1 1990/1999=2
(period: 15 changes made)
. label define periodlab 1 "1980s" 2 "1990s"
. label values period periodlab
. tab period
```

period	Freq.	Percent	Cum.
1980s	7	46.67	46.67
1990s	8	53.33	100.00
Total	15	100.00	

Using the latest version of the `metan` command (Bradburn, Deeks, and Altman 1998), we can examine the effect of magnesium separately, according to time period.

```
. metan dead1 alive1 dead0 alive0, or by(period) label(namevar=trialnam)
```

Study	OR	[95% Conf. Interval]		% Weight
1980s				
Morton	0.436	0.038	5.022	0.09
Rasmussen	0.348	0.154	0.783	0.99
Smith	0.278	0.057	1.357	0.32
Abraham	0.957	0.058	15.773	0.05
Feldstedt	1.250	0.479	3.261	0.35
Schechter	0.090	0.011	0.736	0.42
Ceremuzynski	0.278	0.027	2.883	0.14
Sub-total				
M-H pooled OR	0.437	0.267	0.714	2.36
1990s				
Singh	0.499	0.174	1.426	0.47
Pereira	0.110	0.012	0.967	0.31
Schechter 1	0.130	0.028	0.602	0.57
Golf	0.427	0.127	1.436	0.39
Thogersen	0.452	0.133	1.543	0.37
LIMIT-2	0.741	0.556	0.988	5.04
Schechter 2	0.208	0.067	0.640	0.75
ISIS-4	1.059	0.996	1.127	89.74
Sub-total				
M-H pooled OR	1.020	0.961	1.083	97.64
Overall				
M-H pooled OR	1.007	0.948	1.068	100.00

```
Test(s) of heterogeneity:
             Heterogeneity   degrees of
               statistic      freedom      P    I-squared**
1980s             7.85           6        0.250    23.5%
1990s            30.27           7        0.000    76.9%
Overall          46.61          14        0.000    70.0%
Overall Test for heterogeneity between sub-groups :
                  8.50           1        0.004

** I-squared: the variation in OR attributable to heterogeneity

Significance test(s) of OR=1
1980s          z=  3.31      p = 0.001
1990s          z=  0.66      p = 0.511
Overall        z=  0.22      p = 0.829
```

The by() option of the metafunnel command is used to display separate symbols for the two time periods; the resulting funnel plot is displayed in figure 4.

```
. metafunnel logor selogor, xlab(.05 .1 .25 .5 1 2 4 8 16)
> xscale(log) xtitle(Odds ratio) eform subtitle( )
> ytitle(Standard error of log OR) by(period)
```

As demonstrated by the analysis according to time period, the larger studies were published later. Perhaps more surprisingly, the asymmetry appears to result more from the studies published during the 1990s than from those published during the 1980s.

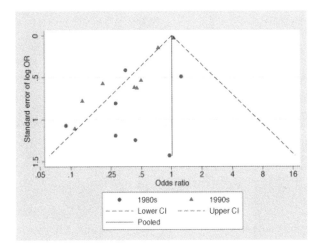

Figure 4. Funnel plot, using data from 15 trials of magnesium therapy following myocardial infarction, grouped according to date of publication.

7 Acknowledgments

Portions of the code for `metafunnel` were originally written by Thomas Steichen, who also gave helpful comments on an early version of the command. We are grateful to Nicholas J. Cox, who provided extensive programming advice.

8 References

Begg, C. B., and M. Mazumdar. 1994. Operating characteristics of a rank correlation test for publication bias. *Biometrics* 50: 1088–1101.

Bradburn, M. J., J. J. Deeks, and D. G. Altman. 1998. sbe24: metan—an alternative meta-analysis command. *Stata Technical Bulletin* 44: 4–15. Reprinted in *Stata Technical Bulletin Reprints*, vol. 8, pp. 86–100. College Station, TX: Stata Press. (Updated article is reprinted in this collection on pp. 3–28.)

Davey Smith, G., and M. Egger. 1994. Who benefits from medical interventions? Treating low risk patients can be a high risk strategy. *British Medical Journal* 308: 72–74.

Deeks, J. J., and D. G. Altman. 2001. Effect measures for meta-analysis of trials with binary outcomes. In *Systematic Reviews in Health Care: Meta-Analysis in Context*, 2nd edition, ed. M. Egger, G. Davey Smith, and D. G. Altman, 313–335. London: BMJ Books.

Egger, M., G. Davey Smith, and D. G. Altman. 2001. *Systematic Reviews in Health Care: Meta-Analysis in Context*, 2nd edition. London: BMJ Books.

Egger, M., G. Davey Smith, M. Schneider, and C. Minder. 1997. Bias in meta-analysis detected by a simple, graphical test. *British Medical Journal* 315: 629–634.

Egger, M., P. Jüni, C. Bartlett, F. Holenstein, and J. A. C. Sterne. 2003. How important are comprehensive literature searches and the assessment of trial quality in systematic reviews? Empirical study. *Health Technology Assessment* 7: 1–68.

Engels, E. A., C. H. Schmid, N. Terrin, I. Olkin, and J. Lau. 2000. Heterogeneity and statistical significance in meta-analysis: An empirical study of 125 meta-analyses. *Statistics in Medicine* 19: 1707–1728.

Glasziou, P. P., and L. M. Irwig. 1995. An evidence based approach to individualising treatment. *British Medical Journal* 311: 1356–1359.

Jüni, P., F. Holenstein, J. A. C. Sterne, C. Bartlett, and M. Egger. 2002. Direction and impact of language bias in meta-analyses of controlled trials: Empirical study. *International Journal of Epidemiology* 31: 115–123.

Poole, C., and S. Greenland. 1999. Random-effects meta-analyses are not always conservative. *American Journal of Epidemiology* 150: 469–475.

Schulz, K. F., I. Chalmers, R. J. Hayes, and D. G. Altman. 1995. Empirical evidence of bias. Dimensions of methodological quality associated with estimates of treatment effects in controlled trials. *Journal of the American Medical Association* 273: 408–412.

Sharp, S. 1998. sbe23: Meta-analysis regression. *Stata Technical Bulletin* 42: 16–22. Reprinted in *Stata Technical Bulletin Reprints*, vol. 7, pp. 148–155. College Station, TX: Stata Press. (Reprinted in this collection on pp. 97–106.)

Steichen, T. J. 1998. sbe19: Tests for publication bias in meta-analysis. *Stata Technical Bulletin* 41: 9–15. Reprinted in *Stata Technical Bulletin Reprints*, vol. 7, pp. 125–133. College Station, TX: Stata Press. (Reprinted in this collection on pp. 151–161.)

Steichen, T. J., M. Egger, and J. A. C. Sterne. 1998. sbe19.1: Tests for publication bias in meta-analysis. *Stata Technical Bulletin* 44: 3–4. Reprinted in *Stata Technical Bulletin Reprints*, vol. 8, pp. 84–85. College Station, TX: Stata Press. (Reprinted in this collection on pp. 162–164.)

Stern, J. M., and R. J. Simes. 1997. Publication bias: Evidence of delayed publication in a cohort study of clinical research projects. *British Medical Journal* 315: 640–645.

Sterne, J. A. C., and M. Egger. 2001. Funnel plots for detecting bias in meta-analysis: Guidelines on choice of axis. *Journal of Clinical Epidemiology* 54: 1046–1055.

Sterne, J. A. C., M. Egger, and G. Davey Smith. 2001. Investigating and dealing with publication and other bias. In *Systematic Reviews in Health Care: Meta-Analysis in Context*, 2nd edition, ed. M. Egger, G. Davey Smith, and D. G. Altman, 189–208. London: BMJ Books.

Sterne, J. A. C., D. Gavaghan, and M. Egger. 2000. Publication and related bias in meta-analysis: Power of statistical tests and prevalence in the literature. *Journal of Clinical Epidemiology* 53: 1119–1129.

Stuck, A. E., A. L. Siu, G. D. Wieland, J. Adams, and L. Z. Rubenstein. 1993. Comprehensive geriatric assessment: A meta-analysis of controlled trials. *Lancet* 342: 1032–1036.

Thompson, S. G., and S. J. Sharp. 1999. Explaining heterogeneity in meta-analysis: A comparison of methods. *Statistics in Medicine* 18: 2693–2708.

Tramer, M. R., D. J. M. Reynolds, R. A. Moore, and H. J. McQuay. 1997. Impact of covert duplicate publication on meta-analysis: A case study. *British Medical Journal* 315: 635–640.

The Stata Journal (2008)
8, Number 2, pp. 242–254

Contour-enhanced funnel plots for meta-analysis

Tom M. Palmer
MRC Centre for Causal Analyses in Translational Epidemiology
Department of Social Medicine
University of Bristol
Bristol, UK
tom.palmer@bristol.ac.uk

Jaime L. Peters
Department of Health Sciences
University of Leicester
Leicester, UK

Alex J. Sutton
Department of Health Sciences
University of Leicester
Leicester, UK

Santiago G. Moreno
Department of Health Sciences
University of Leicester
Leicester, UK

Abstract. Funnel plots are commonly used to investigate publication and related biases in meta-analysis. Although asymmetry in the appearance of a funnel plot is often interpreted as being caused by publication bias, in reality the asymmetry could be due to other factors that cause systematic differences in the results of large and small studies, for example, confounding factors such as differential study quality. Funnel plots can be enhanced by adding contours of statistical significance to aid in interpreting the funnel plot. If studies appear to be missing in areas of low statistical significance, then it is possible that the asymmetry is due to publication bias. If studies appear to be missing in areas of high statistical significance, then publication bias is a less likely cause of the funnel asymmetry. It is proposed that this enhancement to funnel plots should be used routinely for meta-analyses where it is possible that results could be suppressed on the basis of their statistical significance.

Keywords: gr0033, confunnel, funnel plots, meta-analysis, publication bias, small-study effects

1   Introduction

Publication bias is the phenomenon where studies with uninteresting or unfavorable results are less likely to be published than those with more favorable results (Rothstein, Sutton, and Borenstein 2005). If publication bias exists, then the published literature is a biased sample of all studies, and any meta-analysis based on it will be similarly biased.

Funnel plots are commonly used to investigate publication and related biases in meta-analysis (Sterne, Becker, and Egger 2005). They consist of a simple scatterplot of each study's estimate of effect against some measure of its variability, commonly plotted on the x and y axes, respectively (although this goes against the usual convention of plotting the response variable on the y axis). In this way, the studies with the least variable effect sizes appear at the top of the funnel, and the smaller, less precise studies appear at the bottom. In the absence of publication bias, the studies will fan out in a symmetrical funnel shape around the pooled estimate, as variability due to sampling error increases down the y axis. If publication bias is present, then the funnel will appear asymmetric because of the systematic suppression of studies.

A complication in interpreting funnel plots is that funnel asymmetry could be due to factors other than publication bias, such as systematic differences in the results of large and small studies caused by confounding factors such as differential study quality; these differences are sometimes called small-study effects (Sterne and Egger 2001). The aim of the contour-enhanced funnel plot is to aid in disentangling these different causes of funnel asymmetry (Peters et al. 2008).

Funnel plots in Stata were previously described by Sterne and Harbord (2004), and there are several commands available in Stata for drawing funnel plots including `metafunnel`, `funnel` (available with `metan`), and `metabias`. These commands are described in more detail in a frequently asked question about the Stata commands available for meta-analysis; the frequently asked question can be found on Stata's web site at http://www.stata.com/support/faqs/stat/meta.html. In Stata 10, typing `help meta` displays a help file with information about the user-written commands for meta-analysis and tells which are the latest versions.

This article introduces another command for meta-analysis called `confunnel`, which produces contour-enhanced funnel plots. The concept of the contour-enhanced funnel plot is explained in the next section, followed by a description of the command syntax and options. The use of `confunnel` is demonstrated on a well-known meta-analysis example, and the use of the command is also explained in conjunction with some of the other user-written meta-analysis commands.

2 Contour-enhanced funnel plots

There is evidence that, generally, the primary driver for the suppression of studies is the level of statistical significance of study results, with studies that do not attain perceived milestones of statistical significance (i.e., $p < 0.05$ or 0.01) being less likely to be published (Easterbrook et al. 1991; Dickersin 1997; Ioannidis 1998). Despite this, no method has been previously considered to identify the areas of the funnel plot that correspond to different levels of statistical significance, to assess whether any observed asymmetry is likely caused by publication bias.

On a contour-enhanced funnel plot, contours of statistical significance are overlaid on the funnel plot (Peters et al. 2008). Adding contours of statistical significance in this way facilitates the assessment of whether the areas where studies exist are areas of statistical significance and whether the areas where studies are potentially missing correspond to areas of low statistical significance. If studies appear to be missing in areas of low statistical significance, then it is possible that the asymmetry is due to publication bias. Conversely, if the area where studies are perceived to be missing are areas of high statistical significance, then publication bias is a less likely cause of the funnel asymmetry.

There has been discussion as to which is the most informative scale for funnel plots of binary outcome meta-analyses. The consensus is that using the standard error, the variance, or their inverses is most sensible over using an alternative such as sample size (Sterne and Egger 2001; Sterne, Becker, and Egger 2005). Using the standard error on the y axis is easiest to interpret because, in this instance, the contours of statistical significance are linear, which is because they are derived from the Wald statistic for each study's effect estimate. The `confunnel` command has an option to use standard error, inverse standard error, variance, or inverse variance on the y axis.

A meta-analysis of trials investigating magnesium therapy following myocardial infarction is a well-known example in the literature where the presence of publication bias is suspected (Teo et al. 1991; ISIS-4 Collaborative Group 1995; Sterne, Bradburn, and Egger 2001). An initial meta-analysis found that magnesium therapy reduced the risk of mortality; however, a number of larger trials were subsequently published that found no evidence that magnesium therapy reduced the risk of mortality. A standard funnel plot is given for this meta-analysis in figure 1, which was generated by using the `metafunnel` command as shown in the following syntax:

```
. use magnesium
. gen logES = logor
. gen selogES = selogor
. metafunnel logES selogES
```

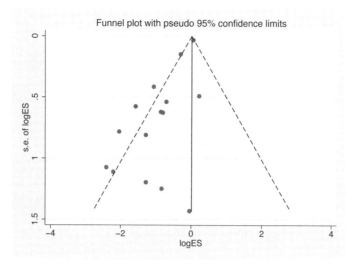

Figure 1. metafunnel funnel plot

When the standard error is used on the y axis of a funnel plot, it is conventional to reverse the axis so that the most precise studies are displayed at the top of the plot.

Figure 1 is compared with the equivalent funnel plot produced by confunnel, shown in figure 2. The addition of the contours of statistical significance makes it easier to assess the proportion of studies published in the meta-analysis at and around statistical significance. The syntax for the default confunnel plot, with the sj scheme, is

```
. confunnel logES selogES
```

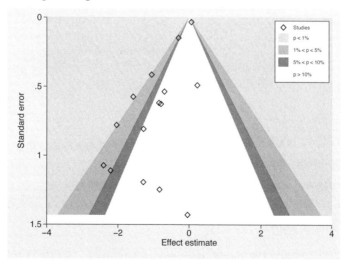

Figure 2. confunnel funnel plot using default options

In both figures 1 and 2, there is a strong suggestion of asymmetry in the funnel, suggesting that studies are missing on the right-hand side of the plot, but figure 2 makes it easier to assess the statistical significance of the hypothetical missing studies. The area where missing studies are perceived includes regions of both low and high statistical significance (i.e., the area crosses over the contours), suggesting studies that showed magnesium to be nonsignificantly and significantly less effective to be missing. Therefore, publication bias cannot be accepted as the only cause of funnel asymmetry if it is believed studies are being suppressed because of a mechanism based on two-sided *p*-values.

It is important to emphasize the differences between the pseudo 95% confidence limits produced by `metafunnel` on figure 1 and the contours of statistical significance produced by `confunnel` on figure 2 (Peters et al. 2008). The pseudo 95% confidence limits illustrate the expected 95% confidence interval about the pooled fixed-effects estimate for the meta-analysis. The pseudo-confidence limits therefore help to assess the extent of between-study heterogeneity in the meta-analysis and the asymmetry on the funnel plot. Unlike the pseudo-confidence limits, the contours of statistical significance are independent of the pooled estimate; therefore, if the pooled estimate is subject to bias, then the contours of significance will not be affected. Also, when the pooled estimate is at the null, the pseudo 95% confidence limits coincide with the two-sided 5% significance contours.

3 The confunnel command

The `confunnel` command plots contour-enhanced funnel plots for study outcome measures in a meta-analysis. Contours of statistical significance from one- or two-sided Wald tests can be plotted using shaded or dashed contour lines. Contours can be plotted along any number of chosen levels of statistical significance; by default, 1%, 5%, and 10% significance contours are plotted. As previously mentioned, `confunnel` has the choice of four *y* axes. The command also has been designed to be flexible, allowing the user to add extra features to the funnel plot.

3.1 Syntax

confunnel *varname1* *varname2* [*if*] [*in*] [, <u>cont</u>ours(*numlist*)

    <u>contc</u>olor(*color*) <u>extraplot</u>(*plots*) <u>functionlow</u>opts(*options*)

    <u>functionupp</u>opts(*options*) <u>legendlabels</u>(*labels*) <u>legendopts</u>(*options*)

    <u>metric</u>(se | invse | var | invvar) <u>one</u>sided(lower | upper)

    <u>scatter</u>opts(*options*) <u>shadedcontours</u> [<u>no</u>]shadedregions <u>solidc</u>ontours

    <u>studylab</u>(*string*) <u>twowayopts</u>(*options*) *twoway_options*]

The first variable, *varname1* is the variable corresponding to the effect estimates, often log odds-ratios, and the second variable, *varname2*, is the variable corresponding to the standard errors of the effect estimates.

3.2 Options

contours(*numlist*) specifies the significance levels of the contours to be plotted; the default is set to 1%, 5%, and 10% significance levels.

contcolor(*color*) specifies the color of the contour lines if shadedcontours is not specified.

extraplot(*plots*) specifies one or multiple additional plots to be overlaid on the funnel plot.

functionlowopts(*options*) and functionuppopts(*options*) pass options to the twoway function commands used to draw the significance contours; for example, the line widths can be changed.

legendlabels(*labels*) specifies labels to appear in the legend for extra elements added to the funnel plot.

legendopts(*options*) passes options to the plot legend.

metric(se | invse | var | invvar) specifies the metric of the y axis of the plot. se, invse, var, and invvar stand for standard error, inverse standard error, variance, and inverse variance, respectively; the default is se.

onesided(lower | upper) can be lower or upper, for lower-tailed or upper-tailed levels of statistical significance, respectively. If unspecified, two-sided significance levels are used to plot the contours.

scatteropts(*options*) specifies any of the options documented in [G] **graph twoway scatter**.

shadedcontours specifies shaded, instead of black, contour lines. Specify this option with the noshadedregions option.

[no]shadedregions suppresses or specifies shaded regions between the contours. This option provides plots that are more similar to those in the original paper by Peters et al. (2008). A plot with shadedregions is the default.

solidcontours specifies solid, instead of dashed, contour lines. Specify this option with both the shadedcontours and the noshadedregions options.

studylab(*string*) specifies the label for the scatter points in the legend. The default is "Studies".

twowayopts(*options*) specifies options passed to the twoway plotting function.

twoway_options are any of the options documented in [G] ***twoway_options***.

4 Use of confunnel

The following subsections use the meta-analysis of magnesium therapy following myocardial infarction.

4.1   Demonstration of some confunnel options

Figure 3 shows the use of the inverse standard error on the y axis; the syntax is as follows:

```
. confunnel logES selogES, metric(invse)
```

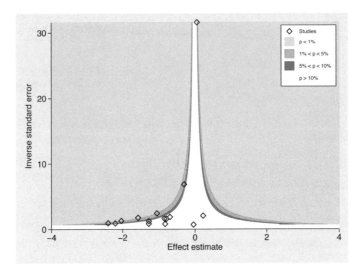

Figure 3. `confunnel` funnel plot using inverse standard error on the y axis

If there is strong evidence that studies are suppressed based on a one-sided (rather than a two-sided) significance test, this can be investigated using the `onesided()` option. This is shown in figure 4 and in the following syntax, which also depicts the contours using lines rather than shaded regions:

```
. confunnel logES selogES, onesided(lower) noshadedregions
```

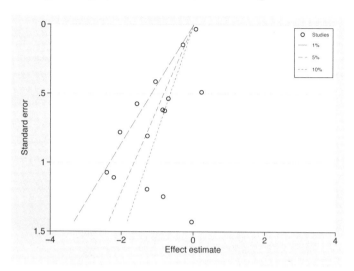

Figure 4. `confunnel` using lower tail one-sided significance regions

Unlike figure 2, in figure 4 (based on one-sided p-values) the area where studies are perceived missing is within the region of low statistical significance. Under this assumption, it is more reasonable to consider publication bias as the potential cause of the funnel asymmetry. In this context, the one-sided assumption implies that studies showing magnesium to be harmful are likely to be suppressed regardless of the significance of the results. Previous methods to address publication bias have made various assumptions about the sidedness of suppression; for example, the trim-and-fill method is one-sided, whereas Egger's regression test is two-sided (Duval and Tweedie 2000; Egger et al. 1997).

Figure 5 shows using variance on the y axis, using the shaded and solid contours options, and labeling the x axis with odds ratios on the funnel plot. The syntax is shown here (`confunnel` was run prior to these commands in order to see where Stata placed the tick marks on the x axis):

(Continued on next page)

```
. local t1 = round(exp(-4)*100)/100
. local t2 = round(exp(-2)*100)/100
. local t3 = exp(0)
. local t4 = round(exp(2)*100)/100
. local t5 = round(exp(4)*100)/100
. confunnel logES selogES, metric(var) twowayopts(xtitle("Odds ratios")
> '"xlabel(-4 "'t1'" -2 "'t2'" 0 "'t3'" 2 "'t4'" 4 "'t5'")"')
```

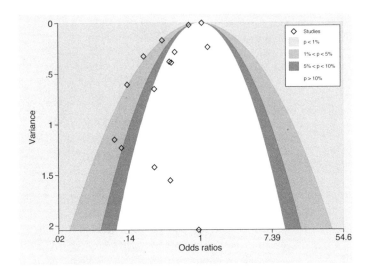

Figure 5. `confunnel` using variance on the y axis

4.2   Use of confunnel with metan, metabias, and metatrim

The `metan` command for meta-analysis (Bradburn, Deeks, and Altman 1998; Harris et al. 2008) can be used to generate the information to display the pooled fixed-effects estimate with its pseudo 95% confidence interval (or, indeed, the pooled random-effects estimate) on the `confunnel` plot; this is shown in figure 6. In this example, because the pooled log odds-ratio was very close to 0, the pseudo 95% confidence interval (for the pooled fixed-effects estimate) almost coincided with the 5% significance contours, which are symmetric about the null hypothesis. The syntax for figure 6 is as follows:

```
. capture drop logES selogES
. metan alive0 dead0 alive1 dead1, or nograph fixed
  (output omitted)
. local fixedlogES = log(r(ES))
. generate logES = log(_ES)
. rename _selogES selogES
. summarize selogES, meanonly
. local semax = r(max)
```

```
. confunnel logES selogES, extraplot(function 'fixedlogES', horizontal
> lc(black) range(0 'semax') || function 'fixedlogES' + x*invnormal(.025),
> horizontal range(0 'semax') lc(black) || function 'fixedlogES' +
> x*invnormal(.975), horizontal range(0 'semax') lc(black))
> legendlabels('"10 "F.E. & 95% C.I.""')
```

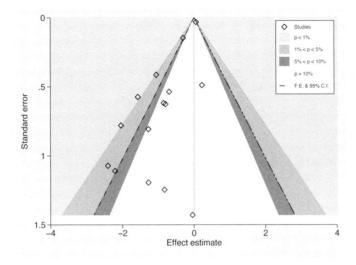

Figure 6. `confunnel` with `metafunnel` features using `metan`

Egger's test investigating possible small-study reporting bias can be represented on the funnel plot by using the information from the `metabias` command (Egger et al. 1997; Steichen 1998); this is shown in figure 7 and in the following syntax:

```
. metabias logES selogES, egger
  (output omitted)
. matrix b = e(b)
. local bias = b[1,2]
. local slope = b[1,1]
. summarize selogES, meanonly
. local semax = r(max)
. metabias alive0 dead0 alive1 dead1, harbord
  (output omitted)
. matrix c = e(b)
. local modbias = c[1,2]
. local modslope = c[1,1]
. confunnel logES selogES, contours(5 10) extraplot(function ('bias'*x + 'slope'),
> horizontal range(0 'semax') lc(black) || function ('modbias'*x + 'modslope'),
> horizontal range(0 'semax') lc(black) lp(dash))
> legendlabels('"7 "Egger" 8 "Harbord""')
```

Also shown on the figure is the modified Egger test using the `metabias` command (Harbord, Harris, and Sterne 2009) because Egger's test has been shown to be biased for binary outcome meta-analyses (Harbord, Egger, and Sterne 2006).

The modified Egger's test is performed on different scales from those of the axes of the funnel plot, but when all trials have a reasonable sample size with small effect estimates, it is not unreasonable to view it on a funnel plot.

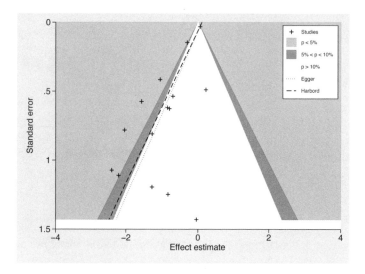

Figure 7. `confunnel` with Egger's and Harbord's regression tests using `metabias`

The trim-and-fill method (Duval and Tweedie 2000) can be applied using `metatrim` (Steichen 2000). In order to demonstrate `confunnel` displaying filled studies, a meta-analysis of the risk of lung cancer from passive smoking is used (Hackshaw, Law, and Wald 1997; Rothstein, Sutton, and Borenstein 2005). Applying the trim-and-fill method, the passive smoking meta-analysis produces seven filled studies, shown in figure 8 and described with the following syntax:

```
. use passivesmoking, clear
. local n = _N
. metan logOR selogOR, nograph
  (output omitted)
. local ES = r(ES)
. summarize selogOR, meanonly
. local semax = r(max)
. metatrim logOR selogOR, save(metatrimdata, replace)
  (output omitted)
. use metatrimdata, clear
. local nfilled = _N - `n'
. metan filled fillse, nograph
  (output omitted)
```

```
. local filledES = r(ES)

. confunnel filled fillse if _n > 'nfilled', contours(5 10)
> extraplot(scatter fillse filled if _n <= 'nfilled', m(T) mc(gs8) ||
> function 'ES', horizontal lc(black) range(0 'semax') ||
> function 'filledES', horizontal lc(gs8) range(0 'semax'))
> legendlabels('"7 "Filled" 8 "F.E." 9 "F.E. filled""')
```

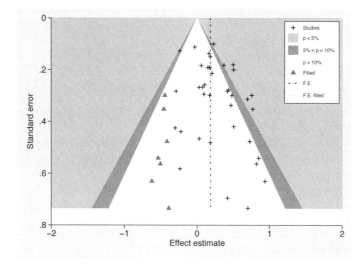

Figure 8. `confunnel` with filled studies from `metatrim`. The vertical dotted line shows the pooled log odds-ratio on the original meta-analysis, while the vertical short dash–dotted line shows the pooled estimate including the filled studies.

It is possible to consider the studies filled by trim and fill as a guide to the likely location of missing studies. With the contours added to the funnel plot containing the filled studies, it is possible to assess the projected significance of the missing studies to determine if it is reasonable to assume such studies could be suppressed by publication bias based on a p-value selection mechanism. In figure 8, trim and fill estimates that seven studies are missing, all of which indicate those exposed to passive smoking are at a reduced risk of lung cancer and all of which are in the region of $p > 0.10$. Hence, it is plausible that publication bias is the cause of the observed asymmetry in this funnel plot.

5 Discussion

The use of the contour-enhanced funnel plot, implemented with the `confunnel` command, is recommended to investigate meta-analyses where it is possible that results could be suppressed on the basis of their statistical significance. In practice, it is suspected that this could include the majority of contexts in which meta-analysis is conducted, certainly in medicine and related disciplines. Exceptions do exist, for example, where noncomparative effect sizes are combined (e.g., in a surgical case series or for incidence or prevalence data); statistical significance will often have no meaning, and in such cases the contours would not be relevant.

An issue with the interpretation of the contour-enhanced funnel plot is that the significance contours can draw the analyst into thinking that the studies should be symmetric about the null hypothesis of the Wald test, because this is the point at which the contours meet when standard error or variance is used on the y axis. But this should be avoided because the studies should form a symmetric funnel shape centered around the true underlying effect size and not the null. Because of this, it can be helpful to plot the meta-analysis pooled estimate for the data on the funnel, although the analyst should be aware that this too may be biased if publication bias is present.

In conclusion, funnel plots are a useful tool in the assessment of systematic differences between the effects in smaller and larger studies in a meta-analysis, regardless of the underlying reason for the differences. Funnel plots can be enhanced by the inclusion of contours of statistical significance, which aid in the interpretation of whether such differences in study estimates in a meta-analysis are most likely to be due to publication bias or other factors.

6   References

Bradburn, M. J., J. J. Deeks, and D. G. Altman. 1998. sbe24: metan—an alternative meta-analysis command. *Stata Technical Bulletin* 44: 4–15. Reprinted in *Stata Technical Bulletin Reprints*, vol. 8, pp. 86–100. College Station, TX: Stata Press. (Updated article is reprinted in this collection on pp. 3–28.)

Dickersin, K. 1997. How important is publication bias? A synthesis of available data. *AIDS Education and Prevention* 9: S15–S21.

Duval, S., and R. Tweedie. 2000. Trim and fill: A simple funnel-plot–based method of testing and adjusting for publication bias in meta-analysis. *Biometrics* 56: 455–463.

Easterbrook, P. J., J. A. Berlin, R. Gopalan, and D. R. Matthews. 1991. Publication bias in clinical research. *Lancet* 337: 867–872.

Egger, M., G. Davey Smith, M. Schneider, and C. Minder. 1997. Bias in meta-analysis detected by a simple, graphical test. *British Medical Journal* 315: 629–634.

Hackshaw, A. K., M. R. Law, and N. J. Wald. 1997. The accumulated evidence on lung cancer and environmental tobacco smoke. *British Medical Journal* 315: 980–988.

Harbord, R. M., M. Egger, and J. A. C. Sterne. 2006. A modified test for small-study effects in meta-analyses of controlled trials with binary endpoints. *Statistics in Medicine* 25: 3443–3457.

Harbord, R. M., R. J. Harris, and J. A. C. Sterne. 2009. sbe19_6: Updated tests for small-study effects in meta-analysis. *Stata Journal*. Forthcoming. (Preprinted in this collection on pp. 138–150.)

Harris, R. J., M. J. Bradburn, J. J. Deeks, R. M. Harbord, D. G. Altman, and J. A. C. Sterne. 2008. metan: fixed- and random-effects meta-analysis. *Stata Journal* 8: 3–28. (Reprinted in this collection on pp. 29–54.)

Ioannidis, J. P. 1998. Effect of the statistical significance of results on the time to completion and publication of randomized efficacy trials. *Journal of the American Medical Association* 279: 281–286.

ISIS-4 Collaborative Group. 1995. ISIS-4: A randomised factorial trial assessing early oral captopril, oral mononitrate, and intravenous magnesium sulphate in 58,050 patients with suspected acute myocardial infarction. *Lancet* 345: 669–685.

Peters, J. L., A. J. Sutton, D. R. Jones, K. R. Abrams, and L. Rushton. 2008. Contour-enhanced meta-analysis funnel plots help distinguish publication bias from other causes of asymmetry. *Journal of Clinical Epidemiology* 61: 991–996.

Rothstein, H. R., A. J. Sutton, and M. Borenstein, ed. 2005. *Publication Bias in Meta-Analysis: Prevention, Assessment and Adjustments*. Chichester, UK: Wiley.

Steichen, T. J. 1998. sbe19: Tests for publication bias in meta-analysis. *Stata Technical Bulletin* 41: 9–15. Reprinted in *Stata Technical Bulletin Reprints*, vol. 7, pp. 125–133. College Station, TX: Stata Press. (Reprinted in this collection on pp. 151–161.)

———. 2000. sbe39: Nonparametric trim and fill analysis of publication bias in meta-analysis. *Stata Technical Bulletin* 57: 8–14. Reprinted in *Stata Technical Bulletin Reprints*, vol. 10, pp. 108–117. College Station, TX: Stata Press. (Reprinted in this collection on pp. 165–177.)

Sterne, J. A. C., B. J. Becker, and M. Egger. 2005. The funnel plot. In *Publication Bias in Meta-Analysis: Prevention, Assessment and Adjustments*, ed. H. R. Rothstein, A. J. Sutton, and M. Borenstein, 75–98. Chichester, UK: Wiley.

Sterne, J. A. C., M. J. Bradburn, and M. Egger. 2001. Meta-analysis in Stata. In *Systematic Reviews in Health Care: Meta-Analysis in Context*, 2nd edition, ed. M. Egger, G. Davey Smith, and D. G. Altman, 347–369. London: BMJ Books.

Sterne, J. A. C., and M. Egger. 2001. Funnel plots for detecting bias in meta-analysis: Guidelines on choice of axis. *Journal of Clinical Epidemiology* 54: 1046–1055.

Sterne, J. A. C., and R. M. Harbord. 2004. Funnel plots in meta-analysis. *Stata Journal* 4: 127–141. (Reprinted in this collection on pp. 109–123.)

Teo, K. K., S. Yusuf, R. Collins, P. H. Held, and R. Peto. 1991. Effects of intravenous magnesium in suspected acute myocardial infarction: Overview of randomised trials. *British Medical Journal* 303: 1499–1503.

The Stata Journal (2009)
Forthcoming

Updated tests for small-study effects in meta-analyses

Roger M. Harbord
Department of Social Medicine
University of Bristol
Bristol, UK
roger.harbord@bristol.ac.uk

Ross J. Harris
Centre for Infections
Health Protection Agency
London, UK
ross.harris@hpa.org.uk

Jonathan A. C. Sterne
Department of Social Medicine
University of Bristol
Bristol, UK

Abstract. This article describes an updated version of the `metabias` command, which provides statistical tests for funnel plot asymmetry. In addition to the previously implemented tests, `metabias` implements two new tests that are recommended in the recently updated *Cochrane Handbook for Systematic Reviews of Interventions* (Higgins and Green 2008). The first new test, proposed by Harbord, Egger, and Sterne (2006, *Statistics in Medicine* 25: 3443–3457), is a modified version of the commonly used test proposed by Egger et al. (1997, *British Medical Journal* 315: 629–634). It regresses $Z/\sqrt{V}$ against $\sqrt{V}$, where Z is the efficient score and V is Fisher's information (the variance of Z under the null hypothesis). The second new test is Peters' test, which is based on a weighted linear regression of the intervention effect estimate on the reciprocal of the sample size. Both of these tests maintain better control of the false-positive rate than the test proposed by Egger at al., while retaining similar power.

Keywords: sbe19_6, metabias, meta-analysis, publication bias, small-study effects, funnel plots

1 Introduction

Publication and related biases in meta-analysis are often examined by visually checking for asymmetry in funnel plots. However, such visual interpretation is inherently subjective. Tests for funnel plot asymmetry (small-study effects [Sterne, Gavaghan, and Egger 2000]) examine whether the association between estimated intervention effects and a measure of study size (such as the standard error of the intervention effect) is greater than might be expected to occur by chance.

This update to the `metabias` command (Steichen 1998; Steichen, Egger, and Sterne 1998) implements two new tests for funnel plot asymmetry that are recommended in the chapter addressing reporting biases (Sterne, Egger, and Moher 2008) in the recent update to the *Cochrane Handbook for Systematic Reviews of Interventions* (Higgins and Green 2008). The modified version of Egger's test (Egger et al. 1997) proposed by Harbord, Egger, and Sterne (2006) still uses linear regression but is based on the efficient score and its variance, Fisher's information. The test proposed by Peters et al. (2006) is based on a weighted linear regression of the intervention effect estimate on the reciprocal of the sample size. These tests address mathematical problems that can occur with the commonly used Egger test and the rank correlation test proposed by Begg and Mazumdar (1994), which was also available in the original version of `metabias`. As with other recently updated meta-analysis commands, the syntax for `metabias` now corresponds to that for the main meta-analysis command, `metan`.

2 Syntax

`metabias` *varlist* [*if*] [*in*], `egger` `harbord` `peters` `begg` [`graph` `nofit` or `rr`
    `level`(*#*) *graph_options*]

As in the `metan` command, *varlist* corresponds to either binary data—in this order: cases and noncases for the experimental group, then cases and noncases for the control group (d_1 h_1 d_0 h_0)—or the intervention effect and its standard error (*theta se_theta*).

The Harbord and Peters tests require binary data. Although the Egger test can be used with binary data, it is recommended only for studies with continuous (numerical) outcome variables and intervention effects measured as mean differences with the format *theta se_theta*.

`by` is allowed with `metabias`; see [D] `by`.

3 Options

`egger`, `harbord`, `peters`, and `begg` specify that the original Egger test, Harbord's modified test, Peters' test, or the rank correlation test proposed by Begg and Mazumdar (1994) be reported, respectively. There is no default; one test must be chosen.

`graph` displays a Galbraith plot (the standard normal deviate of intervention effect estimate against its precision) for the original Egger test or a modified Galbraith plot of $Z/\sqrt{V}$ versus $\sqrt{V}$ for Harbord's modified test. There is no corresponding plot for the Peters or Begg tests.

`nofit` suppresses the fitted regression line and confidence interval around the intercept in the Galbraith plot.

`or` (the default for binary data) uses odds ratios as the effect estimate of interest.

`rr` specifies that risk ratios rather than odds ratios be used. This option is not available for the Peters test.

`level(#)` specifies the confidence level, as a percentage, for confidence intervals. The default is `level(95)` or as set by `set level`; see [U] **20.7 Specifying the width of confidence intervals**.

graph_options are any of the options documented in [G] **graph twoway scatter**. In particular, the options for specifying marker labels are useful.

4   Background

A funnel plot is a simple scatterplot of intervention effect estimates from individual studies against some measure of each study's size or precision (Light and Pillemer 1984; Begg and Berlin 1988; Sterne and Egger 2001). It is common to plot effect estimates on the horizontal axis and the measure of study size on the vertical axis. This is the opposite of the usual convention for twoway plots, in which the outcome (e.g., intervention effect) is plotted on the vertical axis and the covariate (e.g., study size) is plotted on the horizontal axis. The name "funnel plot" arises from the fact that precision of the estimated intervention effect increases as the size of the study increases. Effect estimates from small studies will therefore scatter more widely at the bottom of the graph, with the spread narrowing among larger studies. In the absence of bias, the plot should approximately resemble a symmetrical (inverted) funnel. The `metafunnel` command (Sterne and Harbord 2004) can be used to display funnel plots, while the `confunnel` command (Palmer et al. 2008) can be used to display "contour-enhanced" funnel plots.

Funnel plots are commonly used to assess evidence that the studies included in a meta-analysis are affected by publication bias. If smaller studies without statistically significant effects remain unpublished, this can lead to an asymmetrical appearance of the funnel plot. However, the funnel plot is better seen as a generic means of displaying *small-study effects*—a tendency for the intervention effects estimated in smaller studies to differ from those estimated in larger studies (Sterne, Gavaghan, and Egger 2000). Small-study effects may be due to reporting biases, including publication bias and selective reporting of outcomes (Chan et al. 2004), poor methodological quality leading to spuriously inflated effects in smaller studies, or true heterogeneity (when the size of the intervention effect differs according to study size) (Egger et al. 1997; Sterne, Gavaghan, and Egger 2000). Apparent small-study effects can also be artifactual, because, in some circumstances, sampling variation can lead to an association between the intervention effect and its standard error (Irwig et al. 1998). Finally, small-study effects may be due to chance; this is addressed by statistical tests for funnel plot asymmetry.

For outcomes measured on a continuous (numerical) scale, tests for funnel plot asymmetry are reasonably straightforward. Using an approach proposed by Egger et al. (1997), we can perform a linear regression of the intervention effect estimates on their standard errors, weighting by 1/(variance of the intervention effect estimate). This looks for a straight-line relationship between the intervention effect and its standard error. Under the null hypothesis of no small-study effects, such a line would be vertical on a funnel plot. The greater the association between intervention effect and standard error, the more the slope would move away from vertical. The weighting is important to ensure that the regression estimates are not dominated by the smaller studies. It is mathematically equivalent, however, to a test of zero intercept in an unweighted regression on Galbraith's radial plot (Galbraith 1988) of the standard normal deviate, defined as the effect estimate divided by its standard error, against the precision, defined as the reciprocal of the standard error; and in fact, this method is used in `metabias`. If the regression line on a Galbraith plot is constrained to pass through the origin, its slope gives the summary estimate of fixed-effects meta-analysis as suggested by Galbraith. But if the intercept is estimated, a test of the null hypothesis of zero intercept tests for no association between the effect size and its standard error.

The Egger test has been by far the most widely used and cited approach to testing for funnel plot asymmetry. Unfortunately, there are statistical problems with this approach because the standard error of the log odds-ratio is correlated with the size of the odds ratio due to sampling variability alone, even in the absence of small-study effects (Irwig et al. 1998); see Deeks, Macaskill, and Irwig (2005) for an algebraic explanation of this phenomenon. This can cause funnel plots that were plotted using log odds-ratios (or odds ratios on a log scale) to appear asymmetric and can mean that p-values from the Egger test are too small, leading to false-positive test results. These problems are especially prone to occur when the intervention has a large effect, when there is substantial between-study heterogeneity, when there are few events per study, or when all studies are of similar sizes. Therefore, a number of authors have proposed alternative tests for funnel plot asymmetry. These are reviewed in a new chapter in the recently updated *Cochrane Handbook for Systematic Reviews of Interventions* (Higgins and Green 2008), which also gives guidance on testing for funnel plot asymmetry (Sterne, Egger, and Moher 2008).

4.1 Notation

We shall be primarily concerned with meta-analysis of 2×2 tables, where each study contains an intervention group and a control group, and the outcome is binary. We shall use the notation shown in table 1 for a single 2×2 table, using the letter d to denote those who experience the event of interest and h for those who do not, with subscripts 0 and 1 to indicate the control and intervention groups, respectively. We shall concentrate on the log odds-ratio, ϕ, as the measure of intervention effect, estimated by $\phi = \log(d_1 h_0 / d_0 h_1)$. The usual estimate of the variance of the log odds-ratio is the Woolf formula (Woolf 1955), $\mathrm{Var}(\phi) = 1/d_0 + 1/h_0 + 1/d_1 + 1/h_1$, the square root of which gives the estimated standard error, $\mathrm{SE}(\phi)$.

(Continued on next page)

Table 1. Notation for a single 2×2 table

	Outcome		
	Experienced event d (disease)	Did not experience event h (healthy)	Total
Group 1 (intervention)	d_1	h_1	n_1
Group 2 (control)	d_0	h_0	n_0
Total	d	h	n

The Egger test is based on a two-sided t test of the null hypothesis of zero slope in a linear regression of ϕ against $\mathrm{SE}(\phi)$, weighted by $1/\mathrm{Var}(\phi)$ (Sterne, Gavaghan, and Egger 2000). This is equivalent to a two-sided t test of the null hypothesis of zero intercept in an unweighted linear regression of $\phi/\mathrm{SE}(\phi)$ against $1/\mathrm{SE}(\phi)$, which are the axes used in the Galbraith plot.

4.2 New tests for funnel plot asymmetry

Harbord's modification to Egger's test is based on the component statistics of the score test, namely, the efficient score, Z, and the score variance (Fisher's information), V. Z is the first derivative, and V is minus the second derivative of the log likelihood with respect to ϕ evaluated at $\phi = 0$ (Whitehead and Whitehead 1991; Whitehead 1997). The intercept in a regression of $Z/\sqrt{V}$ against $\sqrt{V}$ is used as a measure of the magnitude of small-study effects, with a two-sided t test of the null hypothesis of zero intercept giving a formal test for small-study effects. This is identical to a test of nonzero slope in a regression of Z/V against $1 = \sqrt{V}$ with weights V. If all marginal totals are considered fixed, V has no sampling error and hence no correlation with Z. If, as seems more realistic, n_0 and n_1 are considered fixed but d and h are not, the correlation remains lower than that between ϕ and its variance as calculated by the Woolf formula, leading to reduced false-positive rates (Harbord, Egger, and Sterne 2006).

Using standard likelihood theory (Whitehead 1997), it can also be shown that when ϕ is small and n is large, $\phi \approx Z/V$ and $\mathrm{Var}(\phi) \approx 1/V$. It follows that the modified test becomes equivalent to the original Egger test when all trials are large and have small effect sizes. A plot of $Z = \sqrt{V}$ against $\sqrt{V}$ is therefore similar to Galbraith's radial plot of $\phi = \mathrm{SE}(\phi)$ against $1/\mathrm{SE}(\phi)$, as noted by Galbraith himself (Galbraith 1988).

When the parameter of interest is the log odds-ratio, ϕ, the efficient score is

$$Z = d_1 - dn_1/n$$

and the score variance evaluated at $\phi = 0$ is

$$V = n_0 n_1 dh/n^2(n-1)$$

The formula for V given above is obtained by using conditional likelihood, conditioning on the marginal totals d and h in table 1. When the parameter of interest is the log risk-ratio, it can be shown by using standard profile likelihood arguments that $Z = (d_1 n - dn_1)/h$ and $V = n_0 n_1 d/(nh)$.

The Peters test is based on a linear regression of ϕ on $1/n$, with weights dh/n. The slope of the regression line is used as a measure of the magnitude of small-study effects, with a two-sided t test of the null hypothesis of zero slope giving a formal test for small-study effects. This is a modification of Macaskill's test (Macaskill, Walter, and Irwig 2001), with the inverse of the total sample size as the independent variable rather than total sample size. The use of the inverse of the total sample size gives more balanced type I error rates in the tail probability areas than where there is no transformation of sample size (Peters et al. 2006). For balanced trials ($n_0 = n_1$), the weights dh/n are proportional to V.

When there is little or no between-trial heterogeneity, the Harbord and Peters tests have false-positive rates close to the nominal level while maintaining similar power to the original linear regression test proposed by Egger et al. (1997) (Harbord, Egger, and Sterne 2006; Peters et al. 2006; Rücker, Schwarzer, and Carpenter 2008).

5 Example

We shall use an example taken from a systematic review of randomized trials of nicotine replacement therapies in smoking cessation (Silagy et al. 2004), restricted to the 51 trials that used chewing gum as the method of delivery.

```
. use nicotinegum
(Nicotine gum for smoking cessation)

. describe

Contains data from nicotinegum.dta
  obs:            51                          Nicotine gum for smoking cessation
  vars:            5                          8 Jan 2009 12:02
  size:           663 (99.9% of memory free)  (_dta has notes)

              storage  display     value
variable name   type   format      label      variable label

trialid        byte    %9.0g
d1             int     %8.0g                   Intervention successes
h1             int     %9.0g                   Intervention failures
d0             int     %8.0g                   Control successes
h0             int     %9.0g                   Control failures

Sorted by:  trialid
```

A standard fixed-effects meta-analysis, with intervention effects measured as odds ratios, suggests that there was a beneficial effect of the intervention (unusually for a medical meta-analysis, the event of interest here, smoking cessation, is good news rather than bad):

```
. metan d1 h1 d0 h0, or nograph
          Study     |   OR    [95% Conf. Interval]    % Weight
------------------+---------------------------------------------------
1                   |  2.253   1.277    3.972        2.18
2                   |  1.850   0.989    3.460        1.98
3                   |  1.039   0.708    1.524        6.96
4                   |  1.416   0.599    3.350        1.21
5                   |  0.977   0.497    1.919        2.33
6                   |  4.773   1.910   11.932        0.70
7                   |  1.761   0.796    3.893        1.26
8                   |  3.159   1.138    8.768        0.69
9                   |  1.533   0.771    3.048        1.83
10                  |  1.385   0.888    2.160        4.55
11                  |  2.949   1.009    8.615        0.61
12                  |  2.293   1.239    4.245        1.92
13                  |  1.234   0.490    3.106        1.12
14                  |  2.624   1.026    6.708        0.87
15                  |  2.035   0.783    5.289        0.82
16                  |  2.822   1.329    5.994        1.13
17                  |  0.869   0.461    1.636        2.82
18                  |  0.887   0.326    2.408        1.10
19                  |  3.404   1.689    6.861        1.18
20                  |  2.170   1.101    4.279        1.59
21                  |  1.412   0.572    3.487        1.08
22                  |  2.029   0.800    5.148        0.97
23                  |  0.955   0.294    3.098        0.77
24                  |  1.250   0.472    3.311        1.00
25                  |  1.847   0.461    7.397        0.41
26                  |  3.327   1.371    8.077        0.76
27                  |  1.434   0.843    2.441        3.16
28                  |  1.333   0.428    4.155        0.72
29                  |  1.235   0.931    1.638       11.86
30                  |  3.142   1.776    5.558        1.84
31                  |  3.522   0.853   14.543        0.28
32                  |  1.168   0.704    1.937        3.81
33                  |  1.511   0.835    2.735        2.45
34                  |  3.824   1.150   12.713        0.39
35                  |  1.165   0.405    3.349        0.85
36                  |  1.345   0.349    5.188        0.50
37                  |  0.483   0.042    5.624        0.26
38                  |  1.713   1.212    2.421        6.33
39                  |  1.393   0.572    3.389        1.09
40                  |  1.844   1.204    2.822        4.30
41                  |  1.460   0.775    2.751        2.18
42                  |  1.269   0.776    2.075        3.84
43                  |  4.110   1.564   10.799        0.59
44                  |  2.082   1.504    2.881        6.57
45                  |  1.714   0.523    5.621        0.57
46                  |  1.294   0.749    2.236        2.98
47                  |  5.313   0.701   40.255        0.20
48                  |  2.703   0.509   14.357        0.25
49                  |  2.124   0.928    4.858        1.07
50                  |  1.760   0.549    5.643        0.58
51                  |  1.460   0.679    3.140        1.49
------------------+---------------------------------------------------
M-H pooled OR       |  1.658   1.515    1.815      100.00
------------------+---------------------------------------------------
```

```
Heterogeneity chi-squared =   62.04 (d.f. = 50) p = 0.118
I-squared (variation in OR attributable to heterogeneity) =   19.4%

Test of OR=1 : z=  10.99 p = 0.000
```

The `metan` command automatically creates the variables _ES, corresponding to the odds ratio, and _selogES, corresponding to the standard error of the log odds-ratio. We can use these to derive variables for input to the `metafunnel` command:

```
. generate logor = log(_ES)
. generate selogor = _selogES
```

We now use `metafunnel` to draw a funnel plot with the log odds-ratio, ϕ, on the horizontal axis and its standard error, $\mathrm{SE}(\phi)$, on the vertical axis. The `egger` option draws a line corresponding to the weighted regression of the log odds-ratio on its standard error that is the basis of Egger's regression test; see figure 1.

```
. metafunnel logor selogor, egger
```

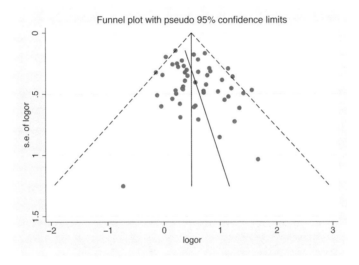

Figure 1. Funnel plot of the log odds-ratio, ϕ, against its standard error, $\mathrm{SE}(\phi)$, including the fitted regression line from the standard regression (Egger) test for small-study effects

The funnel plot appears asymmetric, with smaller studies (those with larger standard errors) tending to have larger (more beneficial) odds ratios. This may suggest publication bias.

We use the `metabias` command to perform a test of small-study effects employing the commonly used Egger test.

```
. metabias d1 h1 d0 h0, egger
Note: data input format tcases tnoncases ccases cnoncases assumed.
Note: odds ratios assumed as effect estimate of interest
Note: peters or harbord tests generally recommended for binary data

Egger's test for small-study effects:
Regress standard normal deviate of intervention
effect estimate against its standard error
Number of studies =  51                           Root MSE       =   1.082
```

Std_Eff	Coef.	Std. Err.	t	P>|t|	[95% Conf. Interval]	
slope	.2832569	.1188368	2.38	0.021	.0444455	.5220683
bias	.7045941	.3566387	1.98	0.054	-.0120982	1.421286

```
Test of H0: no small-study effects         P = 0.054
```

The estimated bias coefficient is 0.705 with a standard error of 0.357, giving a p-value of 0.054. The test thus provides weak evidence for the presence of small-study effects.

The same results can be produced by using the derived variables `logor` and `selogor`:

```
. metabias logor selogor, egger
  (output omitted )
```

We now use Harbord's modified test:

```
. metabias d1 h1 d0 h0, harbord graph
Note: data input format tcases tnoncases ccases cnoncases assumed.
Note: odds ratios assumed as effect estimate of interest

Harbord's modified test for small-study effects:
Regress Z/sqrt(V) on sqrt(V) where Z is efficient score and V is score variance
Number of studies =  51                           Root MSE       =   1.092
```

Z/sqrt(V)	Coef.	Std. Err.	t	P>|t|	[95% Conf. Interval]	
sqrt(V)	.3468707	.126528	2.74	0.009	.0926032	.6011382
bias	.5273137	.3866755	1.36	0.179	-.2497398	1.304367

```
Test of H0: no small-study effects         P = 0.179
```

The estimated intercept is 0.527 with a standard error of 0.387, giving a p-value of 0.179. The modified test thus suggests little evidence for small-study effects. The modified Galbraith plot of $Z/\sqrt{V}$ versus $\sqrt{V}$ is shown in figure 2.

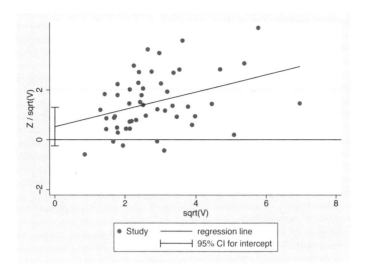

Figure 2. Modified Galbraith plot of $Z/\sqrt{V}$ versus $\sqrt{V}$

Finally, we will use Peters' test for small-study effects:

```
. metabias d1 h1 d0 h0, peters
Note: data input format tcases tnoncases ccases cnoncases assumed.
Note: odds ratios assumed as effect estimate of interest

Peter's test for small-study effects:
Regress intervention effect estimate on 1/Ntot, with weights SF/Ntot
Number of studies =  51                              Root MSE     =  .3897
```

Std_Eff	Coef.	Std. Err.	t	P>|t|	[95% Conf. Interval]
bias	26.20225	14.58572	1.80	0.079	-3.108842 55.51334
constant	.4197904	.0776552	5.41	0.000	.2637364 .5758443

```
Test of H0: no small-study effects          P = 0.079
```

In this example, the p-value from Peters' test is closer to that from Egger's test than it is to the p-value from Harbord's test. These differing results emphasize the importance of selecting a test in advance; picking a test result from among several is strongly discouraged.

(Continued on next page)

6   Saved results

For all tests, the following scalars are returned:

r(N)	number of studies
r(p_bias)	p-value of the bias estimate

For the regression-based tests (Harbord, Peters, and Egger), the following scalars are returned:

r(df_r)	degrees of freedom
r(bias)	estimate of bias (the constant in the regression equation for the Egger and Harbord tests, and the slope for the Peters test)
r(se_bias)	standard error of bias estimate
r(rmse)	root mean squared error of fitted regression model

For Begg's test, the following scalars are returned:

r(score)	Kendall's score ($P-Q$)
r(score_sd)	standard deviation of Kendall's score
r(p_bias_ncc)	p-value for Begg's test (not continuity-corrected)

7   Discussion

We have described how to use the `metabias` command to perform two tests for funnel plot asymmetry. These tests are among those recommended in the *Cochrane Handbook for Systematic Reviews of Interventions* (Higgins and Green 2008) because they reduce the inflation of the false-positive rate (type I error) that can occur for the Egger test, while retaining power compared with alternative tests. `metabias` allows only one test to be specified. Systematic reviewers should ideally specify their chosen test in advance of the analysis and should avoid choosing from among the results of several tests. Although simulation studies comparing the different tests have been reported (Harbord, Egger, and Sterne 2006; Peters et al. 2006; Rücker, Schwarzer, and Carpenter 2008), no test currently has been shown to be superior in all circumstances. A fuller discussion of these issues is available in chapter 10 (Sterne, Egger, and Moher 2008) of the *Cochrane Handbook*.

Tests for funnel plot asymmetry should not be seen as a foolproof method of detecting publication bias or other small-study effects. We recommend that tests for funnel plot asymmetry be used only when there are at least 10 studies included in the meta-analysis. Even then, power may be low. False-positive results may occur in the presence of substantial between-study heterogeneity, and no test performs well when all studies are of a similar size. Although funnel plots, and tests for funnel plot asymmetry, may alert us to a problem that needs considering when interpreting the results of a meta-analysis, they do not provide a solution to this problem.

8 Acknowledgment

We are grateful to Thomas Steichen, who wrote the original version of the `metabias` command and gave us permission to update it.

Some of the guidance in this article is based on the chapter "Addressing reporting biases" (Sterne, Egger, and Moher 2008), published in the new *Cochrane Handbook for Systematic Reviews of Interventions* (Higgins and Green 2008).

9 References

Begg, C. B., and J. A. Berlin. 1988. Publication bias: A problem in interpreting medical data. *Journal of the Royal Statistical Society, Series A* 151: 419–463.

Begg, C. B., and M. Mazumdar. 1994. Operating characteristics of a rank correlation test for publication bias. *Biometrics* 50: 1088–1101.

Chan, A.-W., A. Hróbjartsson, M. T. Haahr, P. C. Gøtzche, and D. G. Altman. 2004. Empirical evidence for selective reporting of outcomes in randomized trials: Comparison of protocols to published articles. *Journal of the American Medical Association* 291: 2457–2465.

Deeks, J. J., P. Macaskill, and L. Irwig. 2005. The performance of tests of publication bias and other sample size effects in systematic reviews of diagnostic test accuracy was assessed. *Journal of Clinical Epidemiology* 58: 882–893.

Egger, M., G. Davey Smith, M. Schneider, and C. Minder. 1997. Bias in meta-analysis detected by a simple, graphical test. *British Medical Journal* 315: 629–634.

Galbraith, R. F. 1988. A note on graphical presentation of estimated odds ratios from several clinical trials. *Statistics in Medicine* 7: 889–894.

Harbord, R. M., M. Egger, and J. A. C. Sterne. 2006. A modified test for small-study effects in meta-analyses of controlled trials with binary endpoints. *Statistics in Medicine* 25: 3443–3457.

Higgins, J. P. T., and S. Green, ed. 2008. *Cochrane Handbook for Systematic Reviews of Interventions.* Chichester, UK: Wiley.

Irwig, L., P. Macaskill, G. Berry, and P. Glasziou. 1998. Bias in meta-analysis detected by a simple, graphical test. Graphical test is itself biased. *British Medical Journal* 316: 470.

Light, R. J., and D. B. Pillemer. 1984. *Summing Up: The Science of Reviewing Research.* Cambridge, MA: Harvard University Press.

Macaskill, P., S. D. Walter, and L. Irwig. 2001. A comparison of methods to detect publication bias in meta-analysis. *Statistics in Medicine* 20: 641–654.

Palmer, T. M., J. L. Peters, A. J. Sutton, and S. G. Moreno. 2008. Contour-enhanced funnel plots for meta-analysis. *Stata Journal* 8: 242–254. (Updated article is reprinted in this collection on pp. 124–137.)

Peters, J. L., A. J. Sutton, D. R. Jones, K. R. Abrams, and L. Rushton. 2006. Comparison of two methods to detect publication bias in meta-analysis. *Journal of the American Medical Association* 295: 676–680.

Rücker, G., G. Schwarzer, and J. Carpenter. 2008. Arcsine test for publication bias in meta-analyses with binary outcomes. *Statistics in Medicine* 27: 746–763.

Silagy, C., T. Lancaster, L. Stead, D. Mant, and G. Fowler. 2004. Nicotine replacement therapy for smoking cessation. *Cochrane Database of Systematic Reviews* 3: CD000146.

Steichen, T. J. 1998. sbe19: Tests for publication bias in meta-analysis. *Stata Technical Bulletin* 41: 9–15. Reprinted in *Stata Technical Bulletin Reprints*, vol. 7, pp. 125–133. College Station, TX: Stata Press. (Reprinted in this collection on pp. 151–161.)

Steichen, T. J., M. Egger, and J. A. C. Sterne. 1998. sbe19.1: Tests for publication bias in meta-analysis. *Stata Technical Bulletin* 44: 3–4. Reprinted in *Stata Technical Bulletin Reprints*, vol. 8, pp. 84–85. College Station, TX: Stata Press. (Reprinted in this collection on pp. 162–164.)

Sterne, J. A. C., and M. Egger. 2001. Funnel plots for detecting bias in meta-analysis: Guidelines on choice of axis. *Journal of Clinical Epidemiology* 54: 1046–1055.

Sterne, J. A. C., M. Egger, and D. Moher. 2008. Addressing reporting biases. In *Cochrane Handbook for Systematic Reviews of Interventions*, ed. J. P. T. Higgins and S. Green, 297–334. Chichester, UK: Wiley.

Sterne, J. A. C., D. Gavaghan, and M. Egger. 2000. Publication and related bias in meta-analysis: Power of statistical tests and prevalence in the literature. *Journal of Clinical Epidemiology* 53: 1119–1129.

Sterne, J. A. C., and R. M. Harbord. 2004. Funnel plots in meta-analysis. *Stata Journal* 4: 127–141. (Reprinted in this collection on pp. 109–123.)

Whitehead, A., and J. Whitehead. 1991. A general parametric approach to the meta-analysis of randomized clinical trials. *Statistics in Medicine* 10: 1665–1677.

Whitehead, J. 1997. *The Design and Analysis of Sequential Clinical Trials*. Rev. 2nd ed. Chichester, UK: Wiley.

Woolf, B. 1955. On estimating the relation between blood group and disease. *Annals of Human Genetics* 19: 251–253.

The Stata Technical Bulletin (1998)
STB-41, pp. 9–15

Tests for publication bias in meta-analysis[1]

Thomas J. Steichen
RJRT
steichen@triad.rr.com

1 Syntax

The syntax of `metabias` is

`metabias` {*theta* {*se_theta* | *var_theta*} | *exp*(*theta*) *ll ul* [*cl*]} [*if*] [*in*] [*,*

 by(*by_var*) graph({begg | egger}) level(*#*) {var | ci} *graph_options*]]

where the syntax construct {*a* | *b* | ...} means choose one and only one of {*a, b, ...*}.

2 Description

`metabias` performs the Begg and Mazumdar (1994) adjusted rank correlation test for publication bias and performs the Egger et al. (1997) regression asymmetry test for publication bias. As options, it provides a funnel graph of the data or the regression asymmetry plot.

The Begg adjusted rank correlation test is a direct statistical analogue of the visual funnel graph. Note that both the test and the funnel graph have low power for detecting publication bias. The Begg and Mazumdar procedure tests for publication bias by determining if there is a significant correlation between the effect estimates and their variances. `metabias` carries out this test by, first, standardizing the effect estimates to stabilize the variances and, second, performing an adjusted rank correlation test based on Kendall's τ.

The Egger et al. regression asymmetry test and the regression asymmetry plot tend to suggest the presence of publication bias more frequently than the Begg approach. The Egger test detects funnel plot asymmetry by determining whether the intercept deviates significantly from zero in a regression of standardized effect estimates against their precision.

The user provides the effect estimate, *theta*, to `metabias` as a log risk-ratio, log odds-ratio, or other direct measure of effect. Along with *theta*, the user supplies a measure of *theta*'s variability (i.e., its standard error, *se_theta*, or its variance, *var_theta*). Alternatively, the user may provide the exponentiated form, *exp*(*theta*), (i.e., a risk ratio or odds ratio) and its confidence interval, (*ll, ul*).

---

1. This article describes the original version of the `metabias` command, which is now obsolete. Syntax for the current version of the command is described in the article by Harbord, Harris, and Sterne (2009).—Ed.

The funnel graph plots *theta* versus *se_theta*. Guide lines to assist in visualizing the funnel are plotted at the variance-weighted (fixed effects) meta-analytic effect estimate and at pseudo confidence interval limits about that effect estimate (i.e., at *theta* $\pm$ $z \times se\_theta$, where z is the standard normal variate for the confidence level specified by option `level()`). Asymmetry on the right of the graph (where studies with high standard error are plotted) may give evidence of publication bias.

The regression asymmetry graph plots the standardized effect estimates, *theta/se_theta*, versus precision, $1/se\_theta$, along with the variance-weighted regression line and the confidence interval about the intercept. Failure of this confidence interval to include zero indicates asymmetry in the funnel plot and may give evidence of publication bias. Guide lines at $x = 0$ and $y = 0$ are plotted to assist in visually determining if zero is in the confidence interval.

`metabias` will perform stratified versions of both the Begg and Mazumdar test and the Egger regression asymmetry test when option `by(by_var)` is specified. Variable *by_var* indicates the categorical variable that defines the strata. The procedure reports results for each strata and for the stratified tests. The graphs, if selected, plot only the combined unstratified data.

3   Options

`by(by_var)` requests that the stratified tests be carried out with strata defined by *by_var*.

`graph(begg)` requests the Begg funnel graph showing the data, the fixed-effects (variance-weighted) meta-analytic effect, and the pseudo confidence interval limits about the meta-analytic effect.

`graph(egger)` requests the Egger regression asymmetry plot showing the standardized effect estimates versus precision, the variance-weighted regression line, and the confidence interval about the intercept.

`level(#)` specifies the confidence level, as a percentage, for confidence intervals. The default is `level(95)` or as set by `set level`; see [U] **20.7 Specifying the width of confidence intervals**.

`var` indicates that *var_theta* was supplied on the command line instead of *se_theta*. Option `ci` should not be specified when option `var` is specified.

`ci` indicates that *exp(theta)* and its confidence interval, (*ll*, *ul*), were supplied on the command line instead of *theta* and *se_theta*. Option `var` should not be specified when option `ci` is specified.

graph_options are those allowed with `graph, twoway`. For `graph(begg)`, the default *graph_options* include `connect(lll.)`, `symbol(iiio)`, and `pen(3552)` for displaying the meta-analytic effect, the pseudo confidence interval limits (two lines), and the data points, respectively. For `graph(egger)`, the default *graph_options* include `connect(.ll)`, `symbol(oid)`, and `pen(233)` for displaying the data points, regression line, and the confidence interval about the intercept, respectively. Setting t2title(.) blanks out the default t2title.

4 Input variables

The effect estimates (and a measure of their variability) can be provided to metabias in any of three ways:

1. The effect estimate and its corresponding standard error (the default method):

   . metabias *theta se_theta* ...

2. the effect estimate and its corresponding variance (note that option var must be specified):

   . metabias *theta var_theta*, var ...

3. the risk (or odds) ratio and its confidence interval (note that option ci must be specified):

   . metabias *exp(theta) ll ul*, ci ...

   where *exp(theta)* is the risk (or odds) ratio, *ll* is the lower limit and *ul* is the upper limit of the risk ratio's confidence interval.

When input method 3) is used, *cl* is an optional input variable that contains the confidence level of the confidence interval defined by *ll* and *ul*:

   . metabias *exp(theta) ll ul cl*, ci ...

If *cl* is not provided, metabias assumes that each confidence interval is at the 95% confidence level. *cl* allows the user to provide the confidence level, by study, when the confidence intervals are not at the default level or are not all at the same level. Values of *cl* can be provided with or without a decimal point. For example, 90 and 0.90 are equivalent and may be mixed (e.g., 90, 0.95, 80, 0.90).

5 Explanation

Meta-analysis has become a popular technique for numerically synthesizing information from published studies. One of the many concerns that must be addressed when performing a meta-analysis is whether selective publication of studies could lead to bias in the meta-analytic conclusions. In particular, if the probability of publication depends on the results of the study—for example, if reporting large or statistically significant findings increase the chance of publication—then the possibility of bias exists.

An initial approach used to assess the likelihood of publication bias was the funnel graph (Light and Pillemer 1984). The funnel graph plotted the outcome measure (effect size) of the component studies against the sample size (a measure of variability).

The approach assumed that all studies in the analysis were estimating the same effect, therefore the estimated effects should be distributed about the unknown true effect level and their spread should be proportional to their variances. This suggested that, when plotted, small studies should be widely spread about the average effect and the spread should narrow as sample sizes increase. If the graph suggested a lack of symmetry about the average effect, especially if small, negative studies were absent, then publication bias was assumed to exist.

Evaluation of a funnel graph was a very subjective process, with bias—or lack of bias—being in the eye of the beholder. Begg and Mazumdar (1994) noted this and observed that the presence of publication bias induced a skewness in the plot and a correlation between the effect sizes and their variances. They proposed that a *formal test for publication bias*, which is implemented in this insert, could be constructed by examining this correlation. The proposed test evaluates the significance of the Kendall's rank correlation between the standardized effect sizes and their variances.

Recently, Egger et al. (1997) proposed an alternative, regression-based test for detecting skewness in the funnel plot and, by extension, for detecting publication bias in the data. This numerical measure of funnel plot asymmetry also constitutes a *formal test for publication bias* and is implemented in this insert. The proposed test evaluates whether the intercept deviates significantly from zero in a regression of standardized effect estimates against their precision. The test is motivated by the observation that, under assumptions of a nonzero underlying effect and a lack of publication bias, 1) small studies would have both a near-zero precision (since precision is predominantly a function of sample size) and a near-zero standardized effect (because of division by a correspondingly large standard error), while 2) large studies would have both a large precision and a large standardized effect (because of division by a small standard error). Therefore the standardized effects would scatter about a regression line (approximately) through the origin that has a slope which estimates both the size and direction of the underlying effect. Under conditions of publication bias and asymmetry in the funnel plot, the subsample of small studies will differ systematically from the subsample of larger studies and the regression line will fail to go through the origin. The size of the intercept provides a measure of asymmetry—the larger the deviation from zero the greater the asymmetry. The direction of the intercept provides information on the form of the bias—a positive intercept indicates that the effect estimated from the smaller studies is greater than the effect estimated from the larger studies. Conversely, a negative intercept indicates that the effect estimated from the smaller studies is less than the effect estimated from the larger studies.

6   Begg's test

This section paraphrases the mathematical development and discussion in the Begg and Mazumdar paper (the paper also includes a detailed examination of the operating characteristics of this test and examples based on medical data).

Let (t_i, v_i), $i = 1, \ldots, k$, be the estimated effect sizes and sample variances from k studies in a meta-analysis. To construct the adjusted rank correlation test, calculate the standardized effect sizes

$$t_i^* = \frac{(t_i - \bar{t})}{(v_i^*)^{1/2}}$$

where

$$\bar{t} = \frac{\sum_{j=1}^{k} t_j v_j^{-1}}{\sum_{j=1}^{k} v_j^{-1}}$$

is the variance-weighted average effect size, and

$$v_i^* = v_i - \left(\sum_{j=1}^{k} v_j^{-1} \right)^{-1}$$

is the variance of $t_i - \bar{t}$.

Correlate the standardized effect sizes, t_i^*, with the sample variances, v_i, using Kendall's rank correlation procedure and examine the p value. A significant correlation is interpreted as providing strong evidence of publication bias.

In their examples, Begg and Mazumdar use the normalized Kendall rank correlation test statistic for data that have no ties, $z = (P - Q)/\{k(k-1)(2k+5)/18\}^{1/2}$, where P is the number of pairs of studies ranked in the same order with respect to t^* and v and Q is the number of pairs ranked in the opposite order. This statistic does not apply a continuity correction. The authors remark that the denominator should be modified if there are tied observations in either t_i^* or v_i but, instead, apparently break ties in their sample data by adding a small constant. The `metabias` procedure implemented in this insert invokes a modification of Stata's `ktau` procedure to calculate the correct statistic, whether ties exist or not, and presents the z and p values with and without the continuity correction.

Begg and Mazumdar (1994) report that the principal determinant of the power of this test is the number of component studies in the meta-analysis (as opposed to the sample sizes of the individual studies). Additionally, the power will increase with a wider range in variance (sample size) and with a smaller underlying effect size. The authors state that the test is fairly powerful for a meta-analysis of 75 component studies, only moderately powerful for one of 25 component studies, and weak when there are few component studies. They advise that "the test must be interpreted with caution in small meta-analyses. In particular, [publication] bias cannot be ruled out if the test is not significant."

A *stratified test* can also be constructed. Let $P_l - Q_l$ be the numerator of the unstratified test statistic for the lth subgroup and d_l be the square of the corresponding denominator (i.e., the variance of $P_l - Q_l$). The stratified test statistic, without continuity correction, is defined as

$$z_s = \frac{\sum_l (P_l - Q_l)}{\left(\sum_l d_l\right)^{1/2}}$$

The `metabias` procedure implemented in this insert calculates the correct stratified statistic, whether ties exist or not, and presents the z_s and p_s values with and without the continuity correction.

Begg and Mazumdar assume that the sampling distribution of t is normal, i.e., $t \sim N(\delta, v_i)$, where δ is the common effect size to be estimated and the v_i are the variances, which depend on the sample sizes of the individual component studies. They argue that the normality assumption is reasonable because t is "invariably a summary estimate of some parameter, and as such will possess an asymptotic normal distribution in most circumstances." The subsequent asymptotic-normality assumption for z_s inherently follows from this argument.

7 Egger's test

This section paraphrases the method development and discussion in the Egger et al. paper. (The paper also provides an empirical evaluation, based on only eight examples from the medical literature, of the ability of the regression asymmetry test to correctly predict whether a meta-analysis of smaller studies will be concordant with the results of a subsequent large trial.)

Let (t_i, v_i), $i = 1, \ldots, k$, be the estimated effect sizes and sample variances from k studies in a meta-analysis. Define the standardized effect size as $t_i^* = t_i/v_i^{1/2}$, the precision as $s^{-1} = 1/v_i^{1/2}$, and the weight as $w_i = 1/v_i$. (In this form of standardization, t^* is a *standard normal deviate* and is designated as such in the Egger paper.) Fit t^* to s^{-1} using standard weighted linear regression with weights w and linear equation: $t^* = \alpha + \beta s^{-1}$. A significant deviation from zero of the estimated intercept, $\widehat{\alpha}$, is interpreted as providing evidence of asymmetry in the funnel plot and of publication bias in the sampled data.

Egger et al. (1997) fit both weighted and unweighted regression lines and select the results of the analysis yielding the intercept with the larger deviation from zero. This insert implements only the weighted analysis.

Egger et al. (1997) do not provide a formal analysis of coverage (i.e., nominal significance level) or power for this test, though they do provide a number of assertions about power. First, they state that "[i]n contrast to the overall test of heterogeneity, the test for funnel plot asymmetry assesses a specific type of heterogeneity and provides a more powerful test in this situation." Second, they state that "[i]n some situations... power is gained by weighting the analysis." Lastly, in a comparison to the Begg and Mazumdar test, they state that "the linear regression approach may be more powerful than the rank correlation test." Egger et al. note, though, that "any analysis of heterogeneity depends on the number of trials included in a meta-analysis, which is generally small, and this limits the statistical power of the test."

Although the paper provides no formal analysis in support of these assertions, an empirical evaluation based on eight examples from the medical literature is reported. This evaluation assessed the ability of the regression asymmetry test to correctly predict whether a meta-analysis of smaller studies will be concordant with the results of a subsequent large trial. For these eight examples, the test detected bias in 3 of 4 cases where a meta-analysis disagreed with a subsequent large trial and indicated no bias in all 4 cases where the meta-analysis agreed with the subsequent large trial. In contrast, the Begg and Mazumdar test was significant for only 1 of the 4 discordant cases (but like Egger's test, for none of the concordant cases). Nonetheless, eight example cases are too few to be statistically convincing and the test remains unvalidated. Further, the lack of coverage analysis leaves open the question of false-positive claims of asymmetry and publication bias. Interestingly, if the Egger's publication bias test is too liberal (a concern that the author of this insert holds) that translates into conservativeness at the meta-analysis level since the bias test will suggest too frequently that caution is needed in interpreting the results of the meta-analysis.

An approximate *stratified test* can be constructed using logic similar to that of Begg and Mazumdar (although Egger et al. did not do so). Let a_l be the intercept from the regression equation for the lth subgroup and v_l^a be the variance of a_l. The stratified test statistic is defined as

$$z_s = \frac{\sum_l a_l / v_l^a}{\left(\sum_l 1/v_l^a\right)^{1/2}}$$

and is assumed to be distributed asymptotically normal. In this form, the stratified estimate is simply the variance-weighted, fixed effect meta-analysis of the intercepts. This stratified test is implemented in this insert.

8   Examples

Begg and Mazumdar illustrated their method with examples from the literature. The first example examined the association between Chlamydia trachomatis and oral contraceptive use derived from 29 case-control studies (Cottingham and Hunter 1992). `metabias` is invoked as follows:

```
. metabias logor varlogor, var graph(egger)
```

Option `var` is used because the data were provided as log odds-ratios and variances and this avoids the, admittedly, small step of generating the standard errors. The optional Egger graph is also requested. `metabias` provides the following analysis:

```
Tests for Publication Bias
Begg's Test
    adj. Kendall's Score (P-Q) =      85
            Std. Dev. of Score =  53.30 (corrected for ties)
            Number of Studies =      29
                            z =    1.59
                    Pr > |z| =   0.111
                            z =    1.58 (continuity corrected)
                    Pr > |z| =   0.115 (continuity corrected)

Egger's Test

  -----------------------------------------------------------------------
  Std_Eff |      Coef.    Std. Err.        t     P>|t|     [95% Conf. Interval]
  --------+--------------------------------------------------------------
    slope |   .5107122    .0266415     19.170    0.000      .4560484    .565376
     bias |   .8016095    .2961195      2.707    0.012      .1940226   1.409196
  -----------------------------------------------------------------------
```

The noncontinuity-corrected test statistic, $z = 1.59$ ($p = 0.111$), differs substantially from that reported by Begg and Mazumdar, $z = 1.76$ ($p = 0.08$). It differs for two reasons: first, the `metabias` procedure corrected the standard deviation of Kendall's score for ties; and second, Begg and Mazumdar apparently carried out their calculation on data that differs slightly from the data they report in their appendix.

The difference in data is apparent when comparing the funnel graph in the published paper to that generated by `metabias`. The published graph suggests that the observation at ($logor, varlogor$) = $(0.41, 0.162)$ incorrectly overlays observation $(0.41, 0.083)$; that it, it was incorrectly entered as $(0.41, 0.083)$. Recalculation of the test statistic with ties broken, and with the data modified to match the published graph, yields the published results.

Begg and Mazumdar report that their p of 0.08 is "strongly suggestive of publication bias." Correction of the data and calculation of the test statistic to account for the ties, as shown above, weakens this conclusion. Application of the continuity correction further weakens the conclusion. Nonetheless, with only 29 component studies, the test is expected to have only moderate power at best, and the existence of publication bias cannot be ruled out.

In contrast, the Egger's bias coefficient, $bias = 0.802$ ($P > |t| = 0.012$), strongly indicates the presence of asymmetry and publication bias. Further, the sign of the coefficient (positive) suggests that small studies overestimate the effect (or, alternatively, that negative and/or nonsignificant small studies are not included in the Cottingham and Hunter dataset). The slope coefficient, 0.511, which is an estimate of *theta* (that in a weak sense might be considered to be adjusted for the effects of publication bias), is smaller than the effects estimated from meta-analysis of these data using either fixed-effects (*theta* = 0.655) or random-effects (*theta* = 0.716). These differences in effect estimates are consistent with those expected when small, negative studies are excluded.

The Egger plot (figure 1), requested via the `graph(egger)` option, graphically shows this test and points out that the analysis is dominated by one large, very precise study. The plot also shows that the data near the origin are systematically elevated.

The Begg funnel graph of the data (figure 2), which could have been selected with the graph(begg) option, provides additional support for this interpretation.

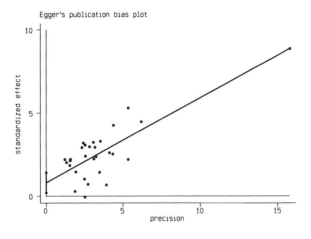

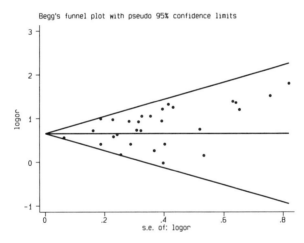

Figure 1. Figure 1 and figure 2

Most of the data points in the Begg plot fall above the meta-analytic effect estimate and there is a visible void in the lower right of the funnel, that is, in the region of low effect and high variance. This is the region where studies most likely to be subject to publication bias would appear. It is notable, though, that since the meta-analytic effect estimate and most of the individual component effect estimates are substantially above zero, the effect of publication bias, if any, would be to inflate the estimate rather than to lead to an incorrect conclusion about the existence of an effect.

Begg and Mazumdar's third example called for the use of the stratified test. These data examined the association between chlorination by-products in drinking water and cancer occurrence, with studies stratified by the site of the cancer (Morris et al. 1992). `metabias` is invoked as follows:

```
. metabias effect variance, var by(site)
```

Use of option `by(site)` informs `metabias` that the stratified tests are to be carried out and that variable `site` is to be used to define the strata. Results are provided in table format, presenting the statistics for each strata and then for the overall stratified tests:

```
Tests for Publication Bias
-----------------------------------------------------------------------------
         |      |    Begg's       Begg's        cont. corr.  |   Egger's
    site |  n | score   s.d.      z      p        z      p   |  bias      p
---------+------+------------------------------------------------+--------------
 Bladder |  7 |   7    6.658    1.05   0.293    0.90   0.368 |  0.07    0.928
   Brain |  2 |   1    1.000    1.00   0.317    0.00   1.000 |  4.71       .
  Breast |  4 |   2    2.944    0.68   0.497    0.34   0.734 |  4.13    0.002
   Colon |  7 |  -1    6.658   -0.15   0.881    0.00   1.000 |  4.36    0.003
 ColoRect |  8 |   0    8.083    0.00   1.000   -0.12   1.000 |  5.33    0.273
 Esophagu |  5 |   4    4.082    0.98   0.327    0.73   0.462 |  1.85    0.456
  Kidney |  4 |   2    2.944    0.68   0.497    0.34   0.734 |  2.31    0.426
   Liver |  4 |   2    2.944    0.68   0.497    0.34   0.734 | -0.78    0.727
    Lung |  5 |   6    4.082    1.47   0.142    1.22   0.221 |  1.06    0.324
 Pancreas |  6 |   5    5.323    0.94   0.348    0.75   0.452 |  1.55    0.001
  Rectum |  6 |   1    5.323    0.19   0.851    0.00   1.000 |  4.39    0.103
 Stomach |  6 |   5    5.323    0.94   0.348    0.75   0.452 |  2.02    0.042
---------+------+------------------------------------------------+--------------
 overall | 64 |  34   17.301    1.97   0.049    1.91   0.056 |  2.51    0.000
-----------------------------------------------------------------------------
```

The Begg and Mazumdar results provide no evidence of publication bias in any of the small site-specific strata, yet the stratified test statistic, $z_s = 1.97$ ($p = 0.049$) provides strong evidence that publication bias exists in the chlorinated drinking water and cancer literature. (These results also differ slightly from those published by Begg and Mazumdar in that the published score for the Pancreas strata is 6, leading to an overall score of 35 and slightly different test statistics for this strata and the overall statistic. Results for all other strata agree.) Again, the Egger test provides a stronger indication of the possible presence of publication bias in this literature. Four site-specific strata (Breast, Colon, Pancreas, and Stomach) reach statistical significance and the p value for the overall test is more significant than that of Begg's test, 0.000 versus 0.049. All but one of the individual bias coefficients are positive, as is the overall bias coefficient, suggesting that the small studies in this Morris et al. dataset are overestimating the effect (or that the negative and/or nonsignificant small studies are not included).

9 Saved results

`metabias` saves the following results:

S_1	number of studies	S_5	Begg's p value, continuity corrected
S_2	Begg's score	S_6	Egger's bias coefficient
S_3	s.d. of Begg's score	S_7	Egger's p value
S_4	Begg's p value	S_8	overall effect (log scale)

10 References

Begg, C. B., and M. Mazumdar. 1994. Operating characteristics of a rank correlation test for publication bias. *Biometrics* 50: 1088–1101.

Cottingham, J., and D. Hunter. 1992. Chlamydia trachomatis and oral contraceptive use: A quantitative review. *Genitourinary Medicine* 68: 209–216.

Egger, M., G. Davey Smith, M. Schneider, and C. Minder. 1997. Bias in meta-analysis detected by a simple, graphical test. *British Medical Journal* 315: 629–634.

Harbord, R. M., R. J. Harris, and J. A. C. Sterne. 2009. sbe19_6: Updated tests for small-study effects in meta-analysis. *Stata Journal.* Forthcoming. (Preprinted in this collection on pp. 138–150.)

Light, R. J., and D. B. Pillemer. 1984. *Summing Up: The Science of Reviewing Research.* Cambridge, MA: Harvard University Press.

Morris, R. D., A. M. Audet, I. F. Angelillo, T. C. Chalmers, and F. Mosteller. 1992. Chlorination, chlorination by-products, and cancer: A meta-analysis. *American Journal of Public Health* 82: 955–963.

The Stata Technical Bulletin (1998)
STB-44, pp. 3–4

Tests for publication bias in meta-analysis[1]

Thomas J. Steichen
RJRT
steichen@triad.rr.com

Matthias Egger
Institute of Social and Preventive Medicine
University of Bern, Switzerland
egger@ispm.unibe.ch

Jonathan A. C. Sterne
Department of Social Medicine
University of Bristol Bristol, UK

1   Modification of the metabias program

This insert documents four changes to the `metabias` program (Steichen 1998). First, the weighted form of the Egger et al. (1997) regression asymmetry test for publication bias has been replaced by the unweighted form. Second, an error has been corrected in the calculation of the asymmetry test p values for individual strata in a stratified analysis. Third, error trapping has been modified to capture or report problem situations more completely and accurately. Fourth, the labeling of the Begg funnel graph has been changed to properly title the axes when the `ci` option is specified. None of these changes affects the program syntax or operation.

The first change was made because, while there is little theoretical justification for the weighted analysis, justification for the unweighted analysis is straightforward. As before, let (t_i, v_i), $i = 1, \ldots, k$, be the estimated effect sizes and sample variances from k studies in a meta-analysis. Egger et al. defined the standardized effect size as $t_i^* = t_i / v_i^{1/2}$, and the precision as $s^{-1} = 1/v_i^{1/2}$. For the unweighted form of the asymmetry test, they fit t^* to s^{-1} using standard linear regression and the equation $t^* = \alpha + \beta s^{-1}$. A significant deviation from zero of the estimated intercept, $\widehat{\alpha}$, is then interpreted as providing evidence of asymmetry in the funnel plot and of publication bias in the sampled data.

Jonathan Sterne (private communication to Matthias Egger) noted that this "unweighted" asymmetry test is merely a reformulation of a standard weighted regression of the original effect sizes, t_i, against their standard errors, $v_i^{1/2}$, where the weights are the usual $1/v_i$. It follows then that the "weighted" asymmetry test is merely a weighted regression of the original effect sizes against their standard errors, but with weights $1/v_i^2$. This form has no obvious theoretical justification.

We note further that the "unweighted" asymmetry test weights the data in a manner consistent with the weighting of the effect sizes in a typical meta-analysis (i.e., both use

---

1. This article describes the original version of the `metabias` command, which is now obsolete. Syntax for the current version of the command is described in the article by Harbord, Harris, and Sterne (2009).—Ed.

the inverse variances). Thus, bias is detected using the same weighting metric as in the meta-analysis.

For these reasons, this insert restricts `metabias` to the unweighted form of the Egger et al. regression asymmetry test for publication bias.

The second change to `metabias` is straightforward. A square root was inadvertently left out of the formula for the p value of the asymmetry test that is calculated for an individual stratum when option `by()` is specified. This formula has been corrected. Users of this program should repeat any stratified analyses they performed with the original program. Please note that unstratified analyses were not affected by this error.

The third change to `metabias` extends the error-trapping capability and reports previously trapped errors more accurately and completely. A noteworthy aspect of this change is the addition of an error trap for the `ci` option. This trap addresses the situation where epidemiological effect estimates and associated error measures are provided to `metabias` as risk (or odds) ratios and corresponding confidence intervals. Unfortunately, if the user failed to specify option `ci` in the previous release, `metabias` assumed that the input was in the default (*theta*, *se_theta*) format and calculated incorrect results. The current release checks for this situation by counting the number of variables on the command line. If more than two variables are specified, `metabias` checks for the presence of option `ci`. If `ci` is not present, `metabias` assumes it was accidentally omitted, displays an appropriate warning message, and proceeds to carry out the analysis as if `ci` had been specified.

Warning: The user should be aware that it remains possible to provide *theta* and its variance, *var_theta*, on the command line without specifying option `var`. This error, unfortunately, cannot be trapped and will result in an incorrect analysis. Though only a limited safeguard, the program now explicitly indicates the data input option specified by the user, or alternatively, warns that the default data input form was assumed.

The fourth change to `metabias` has effect only when options `graph(begg)` and `ci` are specified together. `graph(begg)` requests a funnel graph. Option `ci` indicates that the user provided the effect estimates in their exponentiated form, $exp(theta)$—usually a risk or odds ratio, and provided the variability measures as confidence intervals, (*ll*, *ul*). Since the funnel graph always plots *theta* against its standard error, `metabias` correctly generated *theta* by taking the log of the effect estimate and correctly calculated *se_theta* from the confidence interval. The error was that the axes of the graph were titled using the variable name (or variable label, if available) and did not acknowledge the log transform. This was both confusing and wrong and is corrected in this release. Now when both `graph(begg)` and `ci` are specified, if the variable name for the effect estimate is `RR`, the y axis is titled "log[RR]" and the x axis is titled "s.e. of: log[RR]". If a variable label is provided, it replaces the variable name in these axis titles.

2   References

Egger, M., G. Davey Smith, M. Schneider, and C. Minder. 1997. Bias in meta-analysis detected by a simple, graphical test. *British Medical Journal* 315: 629–634.

Harbord, R. M., R. J. Harris, and J. A. C. Sterne. 2009. sbe19_6: Updated tests for small-study effects in meta-analysis. *Stata Journal.* Forthcoming. (Preprinted in this collection on pp. 138–150.)

Steichen, T. J. 1998. sbe19: Tests for publication bias in meta-analysis. *Stata Technical Bulletin* 41: 9–15. Reprinted in *Stata Technical Bulletin Reprints*, vol. 7, pp. 125–133. College Station, TX: Stata Press. (Reprinted in this collection on pp. 151–161.)

The Stata Technical Bulletin (2000)
STB-57, pp. 8–14

Nonparametric trim and fill analysis of publication bias in meta-analysis

Thomas J. Steichen

RJRT

steichen@triad.rr.com

Abstract. This insert describes `metatrim`, a command implementing the Duval and Tweedie nonparametric "trim and fill" method of accounting for publication bias in meta-analysis. Selective publication of studies, which may lead to bias in estimating the overall meta-analytic effect and in the inferences derived, is of concern when performing a meta-analysis. If publication bias appears to exist, then it is desirable to consider what the unbiased dataset might look like and then to reestimate the overall meta-analytic effect after any apparently "missing" studies are included. Duval and Tweedie's "nonparametric 'trim and fill' method" is an approach designed to meet these objectives.

Keywords: meta-analysis, publication bias, nonparametric, data augmentation

1 Syntax

metatrim {*theta* { *se_theta* | *var_theta* } | *exp(theta) ll ul* [*cl*]} [*if*] [*in*]

    [, {<u>var</u>|ci} <u>ref</u>fect <u>print</u> <u>est</u>imat({<u>r</u>un | <u>l</u>inear | <u>q</u>uadratic}) <u>ef</u>orm

    graph <u>f</u>unnel <u>l</u>evel(*#*) <u>id</u>var(*varname*) <u>sa</u>ve(*filename* [, replace])

    graph_options]

where {*a* | *b* | ...} means choose one and only one of {*a, b, ...*}.

2 Description

`metatrim` performs the Duval and Tweedie (2000) nonparametric "trim and fill" method of accounting for publication bias in meta-analysis. The method, a rank-based data-imputation technique, formalizes the use of funnel plots, estimates the number and outcomes of missing studies, and adjusts the meta-analysis to incorporate the imputed missing data. The authors claim that the method is effective and consistent with other adjustment techniques. As an option, `metatrim` provides a funnel plot of the filled data.

The user provides the effect estimate, *theta*, to `metatrim` as a log risk-ratio, log odds-ratio, or other direct measure of effect. Along with *theta*, the user supplies a measure of *theta*'s variability (that is, its standard error, *se_theta*, or its variance, *var_theta*). Alternatively, the user may provide the exponentiated form, *exp(theta)*, (that is, a risk ratio or odds ratio) and its confidence interval, (*ll*, *ul*).

The funnel plot graphs *theta* versus *se_theta* for the filled data. Imputed observations are indicated by a square around the data symbol. Guide lines to assist in visualizing the center and width of the funnel are plotted at the meta-analytic effect estimate and at pseudo-confidence-interval limits about that effect estimate (that is, at *theta* $\pm z \times$ *se_theta*, where z is the standard normal variate for the confidence level specified by option `level()`).

3   Options

`var` indicates that *var_theta* was supplied on the command line instead of *se_theta*. Option `ci` should not be specified when option `var` is specified.

`ci` indicates that *exp(theta)* and its confidence interval, (*ll*, *ul*), were supplied on the command line instead of *theta* and *se_theta*. Option `var` should not be specified when option `ci` is specified.

`reffect` specifies an analysis based on random-effects meta-analytic estimates. The default is to base calculations on fixed-effects meta-analytic estimates.

`print` requests that the weights used in the filled meta-analysis be listed for each study, together with the individual study estimates and confidence intervals. The studies are labeled by name if the `idvar()` option is specified, or by number otherwise.

`estimat({run | linear | quadratic})` specifies the estimator used to determine the number of points to be trimmed in each iteration. The user is cautioned that the `run` estimator, R_0, is nonrobust to an isolated negative point, and that the `quadratic` estimator, Q_0, may not be defined when the number of points in the dataset is small. The `linear` estimator, L_0, is stable in most situations and is the default.

`eform` requests that the results in the final meta-analysis, and in the `print` option, be reported in exponentiated form. This is useful when the data represent odds ratios or relative risks.

`graph` requests that point estimates and confidence intervals be plotted. The estimate and confidence interval in the graph are derived using fixed- or random-effects meta-analysis, as specified by option `reffect`.

`funnel` requests a filled funnel graph be displayed showing the data, the meta-analytic estimate, and pseudo confidence-interval limits about the meta-analytic estimate. The estimate and confidence interval in the graph are derived using fixed or random-effects meta-analysis, as specified by option `reffect`.

`level(#)` specifies the confidence level, as a percentage, for confidence intervals. The default is `level(95)` or as set by `set level`; see [U] **20.7 Specifying the width of confidence intervals**.

`idvar(`*varname*`)` indicates the character variable used to label the studies.

save(*filename*[, `replace`]) saves the filled data in a separate Stata data file. The *filename* is assumed to have extension `.dta` (an extension should not be provided by the user). If *filename* does not exist, it is created. If *filename* exists, an error will occur unless `replace` is also specified. Only three variables are saved: a study id variable and two variables containing the filled *theta* and *se_theta* values. The study id variable, named `id` in the saved file, is created by `metatrim`; but when option `idvar()` is specified, it is based on that id variable. The filled *theta* and *se_theta* variables are named `filled` and `sefill` in the saved file.

graph_options are those allowed with `graph, twoway`, except `ylabel()`, `symbol()`, `xlog`, `ytick` and `gap` are not recognized by `graph`. For `funnel`, the default *graph_options* include `connect(lll..)`, `symbol(iiioS)`, and `pen(35522)` for displaying the meta-analytic effect, the pseudo confidence interval limits (two lines), and the data points, respectively.

4   Specifying input variables

The individual effect estimates (and a measure of their variability) can be provided to `metatrim` in any of three ways:

1. The effect estimate and its corresponding standard error (the default method):

    . `metatrim` *theta se_theta* ...

2. The effect estimate and its corresponding variance (note that option `var` must be specified):

    . `metatrim` *theta var_theta*, `var` ...

3. The risk (or odds) ratio and its confidence interval (note that option `ci` must be specified):

    . `metatrim` *exp(theta) ll ul*, `ci` ...

    where *exp(theta)* is the risk (or odds) ratio, *ll* is the lower limit and *ul* is the upper limit of the risk ratio's confidence interval.

    When input method 3 is used, *cl* is an optional input variable that contains the confidence level of the confidence interval defined by *ll* and *ul*:

    . `metatrim` *exp(theta) ll ul cl*, `ci` ...

If *cl* is not provided, `metatrim` assumes that a 95% confidence level was reported for each study. *cl* allows the user to combine studies with diverse or non-95% confidence levels by specifying the confidence level for each study not reported at the 95% level. Note that option `level()` does not affect the default confidence level assumed for the individual studies. Values of *cl* can be provided with or without a decimal point. For example, 90 and .90 are equivalent and may be mixed (i.e., 90, .95, 80, .90, etc.). Missing values within *cl* are assumed to indicate a 95% confidence level.

Note that data in binary count format can be converted to the effect format used in `metatrim` by use of program `metan` (Bradburn, Deeks, and Altman 1998). `metan` automatically creates and adds variables for *theta* and *se_theta* to the raw dataset, naming them _ES and _seES. These variables can be provided to `metatrim` using the default input method.

5 Explanation

Meta-analysis is a popular technique for numerically synthesizing information from published studies. One of the many concerns that must be addressed when performing a meta-analysis is whether selective publication of studies could lead to bias in estimating the overall meta-analytic effect and in the inferences derived from the analysis. If publication bias appears to exist, then it is desirable to consider what the unbiased dataset might look like and then to reestimate the overall meta-analytic effect after any apparently "missing" studies are included. Duval and Tweedie's "nonparametric 'trim and fill' method" is designed to meet these objectives and is implemented in this insert.

An early, visual approach used to assess the likelihood of publication bias and to provide a hint of what the unbiased data might look like was the funnel graph (Light and Pillemer 1984). The funnel graph plotted the outcome measure (effect size) of the component studies against the sample size (a measure of variability). The approach assumed that all studies in the analysis were estimating the same effect. Therefore, the effect estimates should be distributed about the unknown true effect level and their spread should be proportional to their variances. This suggested that, when plotted, small studies should be widely spread about the average effect, and the spread should narrow as sample sizes increase, resulting in a symmetric, funnel-shaped graph. If the graph revealed a lack of symmetry about the average effect (especially if small, negative studies appeared to be absent) then publication bias was assumed to exist.

Evaluation of a funnel graph was a very subjective process, with bias—or lack of bias—residing in the eye of the beholder. Begg and Mazumdar (1994) noted this and observed that the presence of publication bias induced skewness in the plot and a correlation between the effect sizes and their variances. They proposed that a formal test of publication bias could be constructed by examining this correlation. More recently, Egger et al. (1997) proposed an alternative, regression-based test for detecting skewness in the funnel plot and, by extension, for detecting publication bias in the data. Their numerical measure of funnel plot asymmetry also constitutes a formal test of publication bias. Stata implementations of both the Begg and Mazumdar procedure and the Egger et al. procedure were provided in `metabias` (Steichen 1998; Steichen, Egger, and Sterne 1998).

However, neither of these procedures provided estimates of the number or characteristics of the missing studies, and neither provided an estimate of the underlying (unbiased) effect. There exist a number of methods to estimate the number of missing studies, model the probability of publication, and provide an estimate of the underlying effect size. Duval and Tweedie list some of these and note that all "are complex and highly computer-intensive to run" and, for these reasons, have failed to find acceptance among meta-analysts. They offer their new method as "a simple technique that seems to meet many of the objections to other methods."

The following sections paraphrase some of the mathematical development and discussion in the Duval and Tweedie paper.

6   Estimators of the number of suppressed studies

Let (Y_j, v_j^2), $j = 1, \ldots, n$, be the estimated effect sizes and within-study variances from n observed studies in a meta-analysis, where all such studies attempt to estimate a common global "effect size" Δ. Define the random-effects model used to combine the Y_j as

$$Y_j = \Delta + \beta_j + \varepsilon_j$$

where $\beta_j \sim N(0, \tau^2)$ accounts for heterogeneity between studies, and $\varepsilon_j \sim N(0, \sigma_j^2)$ is the within-study variability of study j. For a fixed-effects model, assume $\tau^2 = 0$.

Further, in addition to n observed studies, assume that there are k_0 relevant studies that are not observed due to publication bias. Both the value of k_0, that is, the number of unobserved studies, and the effect sizes of these unobserved studies are unknown and must be estimated.

Now, for any collection X_i, $i = 1, \ldots, N$ of random variables, each with a median of zero and sign generated according to an independent set of Bernoulli variables taking values -1 and 1, let r_i denote the rank of $|X_i|$ and

$$W_N^+ = \sum_{X_i > 0} r_i$$

be the sum of the ranks associated with positive X_i. Then W_N^+ has a Wilcoxon distribution.

Assume that among these N random variables, k_0 were suppressed, leaving n observed values. Furthermore, assume that the *suppression has taken place in such a way that the k_0 values of the X_i with the most extreme negative ranks have been suppressed*. (Note: Duval and Tweedie call this their key assumption and present it italicized, as done here, for emphasis. Further, they label the model for an overall set of studies defined in this way as a *suppressed Bernoulli model* and state that it might be expected to lead to a truncated funnel plot.)

Rank again the n observed $|X_i|$ as r_i^* running from 1 to n. Let $\gamma^* \geq 0$ denote the length of the rightmost run of ranks associated with positive values of the observed X_i; that is, if h is the index of the most negative of the X_i and r_h^* is its absolute rank, then $\gamma^* = n - r_h^*$. Define the "trimmed" rank test statistic for the observed n values as

$$T_n = \sum_{X_i > 0} r_i^*$$

Note that though the distributions of γ^* and T_n depend on k_0, the dependence is omitted in this notation. Based on these quantities, define three estimators of k_0, the number of suppressed studies:

$$R_0 = \gamma^* - 1,$$

$$L_0 = \frac{4T_n - n(n+1)}{2n - 1}$$

and

$$Q_0 = n - 1/2 - \sqrt{2n^2 - 4T_n + 1/4}$$

Duval and Tweedie provide the mean and variance of each estimator as follows (the reader should refer to the original paper for the derivation):

$$E(R_0) = k_0, \qquad \mathrm{Var}(R_0) = 2k_0 + 2$$

$$E(L_0) = k_0 - k_0^2/(2n - 1), \qquad \mathrm{Var}(L_0) = 16 \, \mathrm{Var}(T_n)/(2n - 1)^2$$

where

$$\mathrm{Var}(T_n) = \{n(n+1)(2n+1) + 10k_0^3 + 27k_0^2 + 17k_0 - 18nk_0^2 - 18nk_0 + 6n^2 k_0\}/24$$

and

$$E(Q_0) \approx k_0 + \frac{2 \, \mathrm{Var}(T_n)}{\{(n - 1/2)^2 - k_0(2n - k_0 - 1)\}^{3/2}},$$

$$\mathrm{Var}(Q_0) \approx \frac{4 \, \mathrm{Var}(T_n)}{(n - 1/2)^2 - k_0(2n - k_0 - 1)}$$

The authors also report that for n large and k_0 of a smaller order than n, then asymptotically:

$$E(R_0) = k_0$$
$$E(L_0) \sim k_0$$
$$E(Q_0) \sim k_0 + 1/6$$
$$\mathrm{Var}(R_0) = o(n)$$
$$\mathrm{Var}(L_0) \sim n/3$$
$$\mathrm{Var}(Q_0) \sim n/3$$

These results suggest that L_0 and Q_0 should have similar behavior, but the authors report that in practice Q_0 is often larger, sometimes excessively so. They also note that L_0 generally has smaller mean square error than Q_0 when $k_0 \geq n/4 - 2$.

Duval and Tweedie remark that the R_0 *run* estimator is rather conservative and nonrobust to the presence of a relatively isolated negative term at the end of the sequence of ranks. They suggest that the estimators based on T_n seem more robust to such a departure from the suppressed Bernoulli hypothesis. They also note that the Q_0 *quadratic* estimator is defined only when $T_n < n^2/2 + 1/16$, and that simulations show this to be violated quite frequently when the number of studies, n, is small and when the number of suppressed studies, k_0, is large relative to n. These concerns leave the L_0 *linear* estimator as the best all around choice.

Because only whole studies can be trimmed, the estimators are rounded in practice to the nearest nonnegative integer, as follows:

$$R_0^+ = \max(0, R_0)$$

$$L_0^+ = \left\{ \max\left(0, L_0 + \frac{1}{2}\right) \right\}$$

$$Q_0^+ = \left\{ \max\left(0, Q_0 + \frac{1}{2}\right) \right\}$$

where (x) is the integer part of x.

7   The iterative trim and fill algorithm

Because the global "effect size" Δ is unknown, the number and position of any missing studies is correlated with the true value of Δ. Therefore, Duval and Tweedie developed an iterative algorithm to estimate these values simultaneously. The algorithm can be used with any of the three estimators of k_0 defined in the previous section (the `metatrim` program allows the user to specify which one is to be used through the `estimat()` option). Likewise, either a fixed-effects or random-effects meta-analysis model can be used to estimate $\widehat{\Delta}^{(l)}$ within each iteration (l) of the algorithm (the default model in `metatrim` is fixed effects, but random effects is used when option `reffect` is specified). Note that the `meta` program of Sharp and Sterne (1997, 1998) is called by `metatrim` to carry out the meta-analysis calculations.

The algorithm proceeds as follows:

1. Starting with values Y_i, estimate $\widehat{\Delta}^{(1)}$ using the chosen meta-analysis model. Construct an initial set of centered values

$$Y_i^{(1)} = Y_i - \widehat{\Delta}^{(1)}, \quad i = 1, \ldots, n$$

   and estimate $\widehat{k}_0^{(1)}$ using the chosen estimator for k_0 applied to the set of values $Y_i^{(1)}$.

2. Let l be the current step number. Remove $\widehat{k}_0^{(l-1)}$ values from the right end of the original Y_i and estimate $\widehat{\Delta}^{(l)}$ based on this trimmed set of $n - \widehat{k}_0^{(l-1)}$ values: $\{Y_1, \ldots, Y_{n-\widehat{k}_0^{(l-1)}}\}$. Construct the next set of centered values

$$Y_i^{(l)} = Y_i - \widehat{\Delta}^{(l)}, \quad i = 1, \ldots, n$$

   and estimate $\widehat{k}_0^{(l)}$ using the chosen estimator for k_0 applied to the set of values $Y_i^{(l)}$.

3. Increment l and repeat step 2 until an iteration L where $\widehat{k}_0^{(L)} = \widehat{k}_0^{(L-1)}$. Assign this common value to be the estimated value $\widehat{k}_0$. Note that in this iteration it will also be true that $\widehat{\Delta}^{(L)} = \widehat{\Delta}^{(L-1)}$.

4. Augment (that is, "fill") the dataset Y with the $\widehat{k}_0$ imputed symmetric values

$$Y_j^* = 2\widehat{\Delta}^{(L)} - Y_{n-j+1}, \quad j = 1, \ldots, \widehat{k}_0$$

and imputed standard errors

$$\sigma_j^* = \sigma_{n-j+1}, \quad j = 1, \ldots, \widehat{k}_0$$

Estimate the "trimmed and filled" value of Δ using the chosen meta-analysis method applied to the full augmented dataset $\{Y_1, \ldots, Y_n, Y_1^*, \ldots, Y_{\widehat{k}_0}^*\}$.

Conceptually, this algorithm starts with the observed data, iteratively trims (that is, removes) extreme positive studies from the dataset until the remaining studies do not show detectable deviation from symmetry, fills (that is, imputes into the original dataset) studies that are left-side mirrored reflections (about the center of the trimmed data) of the trimmed studies and, finally, repeats the meta-analysis on the filled dataset to get "trimmed and filled" estimates. Each filled study is assigned the same standard error as the trimmed study it reflects in order to maintain symmetry within the filled dataset.

8   Example

The method is illustrated with an example from the literature that examines the association between Chlamydia trachomatis and oral contraceptive use derived from 29 case–control studies (Cottingham and Hunter 1992). Analysis of these data with the publication bias tests of Begg and Mazumdar ($p = 0.115$) and Egger et al. ($p = 0.016$), as provided in `metabias`, suggests that publication bias may affect the data. To examine the potential impact of publication bias on the interpretation of the data, `metatrim` is invoked as follows:

```
. metatrim logor varlogor, reffect funnel var
```

The random-effects model and display of the optional funnel graph are requested via options `reffect` and `funnel`. Option `var` is required because the data were provided as log odds-ratios and variances. By default, the linear estimator, L_0, is used to estimate k_0, as no other estimator was requested. `metatrim` provides the following output:

```
Note: option "var" specified.
Meta-analysis
        | Pooled      95% CI          Asymptotic       No. of
Method  |    Est   Lower   Upper   z_value  p_value    studies
--------+-------------------------------------------------------
Fixed   |  0.655   0.571   0.738   15.359    0.000       29
Random  |  0.716   0.595   0.837   11.594    0.000

Test for heterogeneity: Q= 37.034 on 28 degrees of freedom (p= 0.118)
Moment-based estimate of between studies variance =  0.021

Trimming estimator: Linear
Meta-analysis type: Random-effects model

iteration |  estimate    Tn    # to trim     diff
----------+----------------------------------------
    1     |   0.716     285        5          435
    2     |   0.673     305        6           40
    3     |   0.660     313        7           16
    4     |   0.646     320        7           14
    5     |   0.646     320        7            0

Filled
Meta-analysis
        | Pooled      95% CI          Asymptotic       No. of
Method  |    Est   Lower   Upper   z_value  p_value    studies
--------+-------------------------------------------------------
Fixed   |  0.624   0.542   0.705   14.969    0.000       36
Random  |  0.655   0.531   0.779   10.374    0.000

Test for heterogeneity: Q= 49.412 on 35 degrees of freedom (p= 0.054)
Moment-based estimate of between studies variance =  0.031
```

metatrim first calls program meta to perform and report a standard meta-analysis of the original data, showing both the fixed- and random-effects results. These initial results are always reported as *theta* estimates, regardless of whether the data were provided in exponentiated form.

metatrim next reports the trimming estimator and type of meta-analysis model to be used in the iterative process, then displays results at each iteration. The estimate column shows the value of $\widehat{\Delta}^{(l)}$ at each iteration. As expected, its value at iteration 1 is the same as shown for the random-effects method in the meta-analysis panel, and then decreases in successive iterations as values are trimmed from the data. Column Tn reports the T_n statistic, column # to trim reports the successive estimates $\widehat{k}_0^{(l)}$ and column diff reports the sum of the absolute differences in signed ranks between successive iterations. The algorithm stops when diff is zero.

metatrim finishes with a call to program meta to report an analysis of the trimmed and filled data. Observe that there are now 36 studies, composed of the $n = 29$ observed studies plus the additional $\widehat{k}_0 = 7$ imputed studies. Also note that the estimate of $\widehat{\Delta}$ reported as the random effects pooled estimate for the 36 studies is not the same as the value $\widehat{\Delta}^{(5)}$ shown in the fifth (and final) line of the iteration panel. These values usually differ when the random-effects model is used (because the addition of imputed values change the estimate of τ^2) but are identical always when the fixed-effects model is used.

In summary, `metatrim` adds 7 "missing" studies to the dataset, moving the random-effects summary estimate from $\widehat{\Delta} = 0.716, 95\%$ CI: $(0.595, 0.837)$ to $\widehat{\Delta} = 0.655, 95\%$ CI: $(0.531, 0.779)$. The new estimate, though slightly lower, remains statistically significant; correction for publication bias does not change the overall interpretation of the dataset. Addition of "missing" studies results in an increased variance between studies, the estimate rising from 0.021 to 0.031, and increased evidence of heterogeneity in the dataset, $p = 0.118$ in the observed data versus $p = 0.054$ in the filled data. As expected, when the trimmed and filled dataset is analyzed with the publication bias tests of Begg and Mazumdar (1994) and Egger et al. (1997) (not shown), evidence of publication bias is no longer observed ($p = 0.753$ and $p = 0.690$, respectively).

The funnel plot (figure 1), requested via the `funnel` option, graphically shows the final filled estimate of Δ (as the horizontal line) and the augmented data (as the points), along with pseudo confidence-interval limits intended to assist in visualizing the funnel. The plot indicates the imputed data by a square around the data symbol. The filled dataset is much more symmetric than the original data and the plot shows no evidence of publication bias.

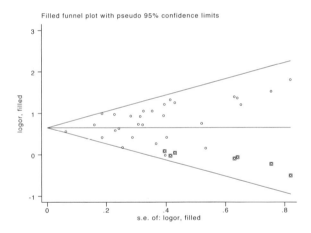

Figure 1. Funnel plot for analysis of Cottingham and Hunter data

Additional options that can be specified include `print` to show the weights, study estimates and confidence intervals for the filled dataset, `eform` to request that the results be reported in exponentiated form in the final meta-analysis and in the `print` option be reported in exponentiated form (this is useful when the data represent odds ratios or relative risks), `graph` to graphically display the study estimates and confidence intervals for the filled dataset, and `save(`*filename*`)` to save the filled data in a separate Stata dataset.

9 Remarks

The Duval and Tweedie method is based on the observation that an unbiased selection of studies that estimate the same thing should be symmetric about the underlying common effect (at least within sampling error). This implies an expectation that the number of studies, and the magnitudes of those studies, should also be roughly equivalent both above and below the common effect value. It is, therefore, reasonable to apply a nonparametric approach to test these assumptions and to adjust the data until the assumptions are met. The price of the nonparametric approach is, of course, lower power (and a concomitant expectation that one may under-adjust the data).

Duval and Tweedie use the symmetry argument in a somewhat roundabout way, choosing to first trim extreme positive studies until the remaining studies meet symmetry requirements. This makes sense when the studies are subject only to publication bias, since trimming should preferably toss out the low-weight, but extreme studies. Nonetheless, if other biases affect the data, in particular if there is a study that is high-weight and extremely positive relative to the remainder of the studies, then the method could fail to function properly. The user must remain alert to such possibilities.

Duval and Tweedie's final step—giving imputed reflections of the trimmed studies—has no effect on the final trimmed point estimate in a fixed effects analysis but does cause the confidence interval of the estimate to be smaller than that from the trimmed or original data. One could question whether this "increased" confidence is warranted.

The random-effects situation is more complex, as both the trimmed point estimate and confidence interval width are affected by filling, with a tendency for the filled data to yield a point estimate between the values from the original and trimmed data. When the random-effects model is used, the confidence interval of the filled data is typically smaller than that of either the trimmed or original data.

Experimentation suggests that the Duval and Tweedie method trims more studies than may be expected; but because of the increase in precision induced by the imputation of studies during filling, changes in the "significance" of the results occur less often than expected. Thus the two operations (trimming, which reduces the point estimate, and filling, which increases the precision) seem to counter each other.

Another phenomenon noted is a tendency for the heterogeneity of the filled data to be greater than that of the original data. This suggests that the most likely studies to be trimmed and filled are those that are most responsible for heterogeneity. The generality of this phenomenon and its impact on the analysis have not been investigated.

Duval and Tweedie provide a reasonable development based on accepted statistics; nonetheless, the number and the magnitude of the assumptions required by the method are substantial. If the underlying assumptions hold in a given dataset, then, as with many methods, it will tend to under- rather than overcorrect. This is an acceptable situation in my view (whereas "overcorrection" of publication bias would be a critical flaw).

This author presents the program as an *experimental* tool only. Users must assess for themselves both the amount of correction provided and the reasonableness of that correction. Other tools to assess publication bias issues should be used in tandem. `metatrim` should be treated as merely one of an arsenal of methods needed to fully assess a meta-analysis.

10   Saved results

`metatrim` does not save values in the system S_# macros, nor does it return results in `r()`.

11   Note

The command `meta` (Sharp and Sterne 1997, 1998) should be installed before running `metatrim`.

12   References

Begg, C. B., and M. Mazumdar. 1994. Operating characteristics of a rank correlation test for publication bias. *Biometrics* 50: 1088–1101.

Bradburn, M. J., J. J. Deeks, and D. G. Altman. 1998. sbe24: metan—an alternative meta-analysis command. *Stata Technical Bulletin* 44: 4–15. Reprinted in *Stata Technical Bulletin Reprints*, vol. 8, pp. 86–100. College Station, TX: Stata Press. (Updated article is reprinted in this collection on pp. 3–28.)

Cottingham, J., and D. Hunter. 1992. Chlamydia trachomatis and oral contraceptive use: A quantitative review. *Genitourinary Medicine* 68: 209–216.

Duval, S., and R. Tweedie. 2000. A nonparametric "trim and fill" method of accounting for publication bias in meta-analysis. *Journal of the American Statistical Association* 95: 89–98.

Egger, M., G. Davey Smith, M. Schneider, and C. Minder. 1997. Bias in meta-analysis detected by a simple, graphical test. *British Medical Journal* 315: 629–634.

Light, R. J., and D. B. Pillemer. 1984. *Summing Up: The Science of Reviewing Research.* Cambridge, MA: Harvard University Press.

Sharp, S., and J. A. C. Sterne. 1997. sbe16: Meta-analysis. *Stata Technical Bulletin* 38: 9–14. Reprinted in *Stata Technical Bulletin Reprints*, vol. 7, pp. 100–106. College Station, TX: Stata Press.[1]

---

1. The original command to perform meta-analysis was `meta`, documented in the sbe16 articles; `meta` is now `metan`. `metan` is described in an updated article, sbe24, on pages 3–28 of this collection.—Ed.

————. 1998. sbe16.1: New syntax and output for the meta-analysis command. *Stata Technical Bulletin* 42: 6–8. Reprinted in *Stata Technical Bulletin Reprints*, vol. 7, pp. 106–108. College Station, TX: Stata Press.[1]

Steichen, T. J. 1998. sbe19: Tests for publication bias in meta-analysis. *Stata Technical Bulletin* 41: 9–15. Reprinted in *Stata Technical Bulletin Reprints*, vol. 7, pp. 125–133. College Station, TX: Stata Press. (Reprinted in this collection on pp. 151–161.)

Steichen, T. J., M. Egger, and J. A. C. Sterne. 1998. sbe19.1: Tests for publication bias in meta-analysis. *Stata Technical Bulletin* 44: 3–4. Reprinted in *Stata Technical Bulletin Reprints*, vol. 8, pp. 84–85. College Station, TX: Stata Press. (Reprinted in this collection on pp. 162–164.)

Part 4

Advanced methods: metandi, glst, metamiss, and mvmeta

In its simplest form, meta-analysis is a weighted average of the effect estimates from different studies. However, more advanced methods require complex computational routines, which are greatly facilitated by the sophisticated estimation procedures available to users writing Stata commands. The last set of articles in this collection describes advanced meta-analysis commands.

There is increasing interest, in the medical literature, in systematic reviews and meta-analyses of studies that estimate the accuracy of diagnostic tests. Such meta-analyses are inherently bivariate, because of the trade off between sensitivity and specificity. Harbord et al. (2007) noted that two apparently different proposed methods for meta-analyses of test accuracy studies (Rutter and Gatsonis 2001; Reitsma et al. 2005) were identical in many circumstances. These methods are implemented in the `metandi` command.

The `glst` command is extremely useful to those wanting to conduct meta-analyses of observational data reported as dose–response associations (for example, associations between alcohol consumption and cardiovascular mortality, or between consumption of beta-carotene and lung cancer). Individual papers often report such associations as risk ratios or odds ratios comparing two or more exposure levels with a baseline category. The `glst` command uses the method of Greenland and Longnecker (1992) to convert such data to an estimate of the dose–response relationship—the risk ratio (or odds ratio) per unit increase in exposure. `glst` can also estimate a summary linear trend across multiple studies. Alternatively, a dose–response meta-analysis can be derived by using the log of the dose–response estimate, with its standard error, as input to the `metan` command.

The collection finishes with two recent articles describing advanced routines. The `metamiss` command performs meta-analysis with binary outcomes in which results from some studies are missing (White and Higgins 2009). Results adjusted for bias are obtained via assumptions about the informative missingness odds ratio. Multivariate random-effects meta-analysis—implemented in the `mvmeta` command (White 2009) can be used in a variety of circumstances, including modeling each outcome separately in a clinical trial, exploring treatment effects on both disease outcome and costs, and examining the shape of an association of a quantitative exposure with a disease outcome.

1   References

Greenland, S., and M. P. Longnecker. 1992. Methods for trend estimation from summarized dose–reponse data, with applications to meta-analysis. *American Journal of Epidemiology* 135: 1301–1309.

Harbord, R. M., J. J. Deeks, M. Egger, P. Whiting, and J. A. C. Sterne. 2007. A unification of models for meta-analysis of diagnostic accuracy studies. *Biostatistics* 8: 239–251.

Reitsma, J. B., A. S. Glas, A. W. S. Rutjes, R. J. P. M. Scholten, P. M. Bossuyt, and A. H. Zwinderman. 2005. Bivariate analysis of sensitivity and specificity produces informative summary measures in diagnostic reviews. *Journal of Clinical Epidemiology* 58: 982–990.

Rutter, C. M., and C. A. Gatsonis. 2001. A hierarchical regression approach to meta-analysis of diagnostic test accuracy evaluations. *Statistics in Medicine* 20: 2865–2884.

White, I. R. 2009. Multivariate random-effects meta-analysis. *Stata Journal.* Forthcoming. (Preprinted in this collection on pp. 231–247.)

White, I. R., and J. P. T. Higgins. 2009. Meta-analysis with missing data. *Stata Journal.* Forthcoming. (Preprinted in this collection on pp. 218–230.)

metandi: Meta-analysis of diagnostic accuracy using hierarchical logistic regression

Roger M. Harbord
Department of Social Medicine
University of Bristol
Bristol, UK
roger.harbord@bristol.ac.uk

Penny Whiting
Department of Social Medicine
University of Bristol
Bristol, UK

Abstract. Meta-analysis of diagnostic test accuracy presents many challenges. Even in the simplest case, when the data are summarized by a 2×2 table from each study, a statistically rigorous analysis requires hierarchical (multilevel) models that respect the binomial data structure, such as hierarchical logistic regression. We present a Stata package, `metandi`, to facilitate the fitting of such models in Stata. The commands display the results in two alternative parameterizations and produce a customizable plot. `metandi` requires either Stata 10 or above (which has the new command `xtmelogit`), or Stata 8.2 or above with `gllamm` installed.

Keywords: metandi, metandiplot, diagnosis, meta-analysis, sensitivity and specificity, hierarchical models, generalized mixed models, gllamm, xtmelogit, receiver operating characteristic (ROC), summary ROC, hierarchical summary ROC

1 Introduction

There are several existing user-written commands in Stata that are intended primarily for meta-analysis (see Sterne et al. [2007] for an overview). There is increasing interest in systematic reviews and meta-analyses of data from diagnostic accuracy studies (Deeks 2001b; Devillé et al. 2002; Tatsioni et al. 2005; Gluud and Gluud 2005; Mallett et al. 2006; Gatsonis and Paliwal 2006), which presents many additional challenges compared to more traditional meta-analysis applications, such as controlled trials. In particular, diagnostic accuracy cannot be adequately summarized by one measure; two measures are typically used, most often sensitivity and specificity or, alternatively, positive and negative likelihood ratios, and the two are correlated (Deeks 2001a). Meta-analysis of diagnostic accuracy therefore requires different and more complex methods than traditional meta-analysis applications, even in the simplest situation where the data from each primary study are summarized as a 2×2 table of test results against true disease status, both of which have been dichotomized. In addition, substantial between-study heterogeneity is commonplace, and the models must account for this (Lijmer, Bossuyt, and Heisterkamp 2002).

Several methods of meta-analyzing diagnostic accuracy data have been proposed, of which two are statistically rigorous: the hierarchical summary receiver operating characteristic (HSROC) model (Rutter and Gatsonis 2001) and the bivariate model (Reitsma et al. 2005). In the absence of covariates, these turn out to be different parameterizations of the same model (Harbord et al. 2007; Arends et al. 2008).

The bivariate model can be fit in Stata by using the user-written `gllamm` command, as pointed out by Coveney (2004). In Stata 10, the same model can be fit considerably faster by using the new `xtmelogit` command. In either case, however, some data preparation is required, the syntax is complex (particularly for `gllamm`), and the output is not easy to interpret.

In this article, we present a new Stata command, `metandi`, to facilitate the fitting of these hierarchical logistic regression models for meta-analysis of diagnostic test accuracy. The `metandi` command fits the model and displays the estimates in both the HSROC and bivariate parameterizations. `metandi` also displays some familiar summary measures (sensitivity and specificity, positive and negative likelihood ratios, and the diagnostic odds ratio). However, these simple summary measures fail to describe the expected trade-off between sensitivity and specificity, which is best illustrated graphically. We have therefore included a command, `metandiplot`, to simplify the plotting of graphical summaries of the fitted model, namely, the summary receiver operating characteristic (SROC) curve and the prediction region, and also to plot the summary point and its confidence region.

The name `metandi` was chosen to indicate that, like `metan` (Bradburn, Deeks, and Altman 1998), `metandi` takes the cell counts of 2×2 tables as input but is designed for meta-analysis of diagnostic accuracy.

`metandi` is not intended to provide a comprehensive package for diagnostic meta-analysis by itself; other plots are also useful, such as forest plots showing within-study estimates and confidence intervals for sensitivity and specificity separately (Deeks 2001b).

Section 2 of this article introduces an example dataset, which we will use to illustrate the commands. Section 3 then gives some background on methods and models that have been proposed for meta-analysis of diagnostic accuracy. Sections 4 and 5 illustrate the output of `metandi` and `metandiplot` on the example dataset. Section 6, which assumes somewhat greater knowledge of both statistics and Stata, gives examples of the use of `predict` after `metandi` for model checking and identification of influential studies. Finally, sections 7 and 8, which are intended mainly as reference material, detail the formal syntax of the commands, and the methods and formulas used.

2 Example: Lymphangiography for diagnosis of lymph node metastasis

We shall illustrate the use of the `metandi` package on data from 17 studies of lymphangiography for the diagnosis of lymph node metastasis in women with cervical cancer. Lymphangiography is one of three imaging techniques in the meta-analysis of Scheidler et al. (1997), and these data have been frequently used as an example for methodological papers on meta-analysis of diagnostic accuracy (Rutter and Gatsonis 2001; Macaskill 2004; Reitsma et al. 2005; Harbord et al. 2007). These data are provided in the auxiliary file `scheidler_LAG.dta`. The total number of patients in each study ranges from 21 to 300. There is one observation in the dataset for each study. The data needed for meta-analysis consist of the number of true positives (tp), false positives (fp), false negatives (fn), and true negatives (tn).

Figure 1 shows a SROC plot of these data, generated by the official Stata commands given below. An SROC plot is similar to a conventional ROC plot (see, e.g., [R] **roc**) in that it plots sensitivity (true-positive rate) against specificity (true-negative rate), but here each symbol represents a different study rather than a different threshold within the same study. It therefore makes no sense to connect the points with a line, but it can be useful to indicate the size of each study by the symbol size. (It might be preferable to use an ellipse or rectangle to separately indicate the number of people with [tp + fn] and without [tn + fp] the disease of interest, but this is hard to achieve within the current Stata graphics system.) By convention, the specificity is plotted on a reversed scale (or equivalently, the false-positive rate is plotted on a conventional scale).

```
. use scheidler_LAG
(Lymphangiography for diagnosing lymph node metastases)
. generate sens = tp/(tp+fn)
. generate spec = tn/(tn+fp)
. label variable sens "Sensitivity"
. label variable spec "Specificity"
. local opts "xscale(reverse) xla(0(.2)1) yla(0(.2)1, nogrid) aspect(1) nodraw"
. scatter sens spec [fw=tp+fp+fn+tn], m(Oh) 'opts' name(sroccirc)
. scatter sens spec, mlabel(studyid) m(i) mlabpos(0) 'opts' name(sroclab)
. graph combine sroccirc sroclab, xsize(4.5) scale(*1.5)
```

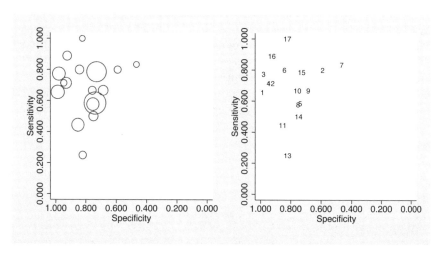

Figure 1. SROC plot of the lymphangiography data. Left panel: Studies indicated by circles sized according to the total number of individuals in each study. Right panel: Studies indicated by study ID numbers.

3   Models for meta-analysis of diagnostic accuracy

Several statistical methods for meta-analysis of data from diagnostic test accuracy studies have been proposed that account for the correlation between sensitivity and specificity (Moses, Shapiro, and Littenberg 1993; Rutter and Gatsonis 2001; Reitsma et al. 2005).

Moses, Shapiro, and Littenberg (1993) proposed a method of generating an SROC curve by using simple linear regression. This method has frequently been used, but the assumptions of simple linear regression are not met, and the method is therefore approximate. There is also uncertainty as to the most appropriate weighting of the regression (Walter 2002; Rutter and Gatsonis 2001).

Two more-complex but statistically rigorous approaches have been proposed that overcome the limitations of the linear regression method: the HSROC model (Rutter and Gatsonia 2001) and the bivariate model (Reitsma et al. 2005). Both approaches are based on hierarchical models, i.e., both approaches involve statistical distributions at two levels. At the lower level, they model the cell counts in the 2×2 tables by using binomial distributions and logistic (log-odds) transformations of proportions. Although their motivation is distinct and they allow covariates to be added to the models in different ways, it has been shown that the two models are equivalent when no covariates are fit, as well as in certain models including covariates (Harbord et al. 2007; Arends et al. 2008).

3.1   HSROC model

The HSROC model (Rutter and Gatsonis 2001) assumes that there is an underlying ROC curve in each study with parameters α and β that characterize the accuracy and asymmetry of the curve. The 2×2 table for each study then arises from dichotomizing at a positivity threshold, θ. The parameters α and θ are assumed to vary between studies; both are assumed to have normal distributions as in conventional random-effects meta-analysis. The accuracy parameter has a mean of Λ (capital lambda) and a variance of σ_α^2, while the positivity parameter θ has a mean of Θ (capital theta) and a variance of σ_θ^2. Because estimation of the shape parameter, β, requires information from more than one study, it is assumed constant across studies. When no covariates are included in an HSROC model, there are therefore five parameters: Λ, Θ, β, σ_α^2, and σ_θ^2.

3.2   Bivariate model

The bivariate model (Reitsma et al. 2005) models the sensitivity and specificity more directly. It assumes that their logit (log-odds) transforms have a bivariate normal distribution between studies. The logit-transformed sensitivities are assumed to have a mean of μ_A and a variance of σ_A^2, while the logit-transformed specificities have a mean of μ_B and a variance of σ_B^2. The trade-off between sensitivity and specificity is allowed for by also including a correlation, ρ_{AB}, that is expected to be negative. The bivariate model, like the HSROC model, therefore has five parameters when no covariates are included: μ_A, μ_B, σ_A^2, σ_B^2, and ρ_{AB}.

4   metandi output

The output from running `metandi` on the lymphangiography data is shown below (the `nolog` option suppresses the iteration log and is used here merely to save space):

```
. use scheidler_LAG, clear
(Lymphangiography for diagnosing lymph node metastases)

. metandi tp fp fn tn, nolog
True  positives: tp                    False positives: fp
False negatives: fn                    True  negatives: tn

Meta-analysis of diagnostic accuracy

Log likelihood    = -91.391372                 Number of studies =      17
```

	Coef.	Std. Err.	z	P>|z|	[95% Conf. Interval]	
Bivariate						
E(logitSe)	.7266321	.1544626			.4238909	1.029373
E(logitSp)	1.638955	.2505372			1.147911	2.129999
Var(logitSe)	.1249622	.1306738			.0160943	.9702552
Var(logitSp)	.8232703	.4055446			.3135009	2.161952
Corr(logits)	.2387873	.4557706			-.6067877	.8308258
HSROC						
Lambda	2.187142	.3086554			1.582189	2.792096
Theta	.0705698	.3271092			-.5705525	.7116921
beta	.9426366	.5764601	1.64	0.102	-.1872044	2.072478
s2alpha	.7946708	.5114529			.2250873	2.805586
s2theta	.1220778	.1082908			.0214569	.6945553
Summary pt.						
Se	.6740658	.0339356			.6044139	.7367944
Sp	.8373927	.0341147			.7591292	.8937849
DOR	10.65029	3.296352			5.806411	19.53509
LR+	4.145361	.9181013			2.685598	6.398582
LR-	.389225	.0452324			.3099427	.4887875
1/LR-	2.569208	.2985712			2.045879	3.226402

```
Covariance between estimates of E(logitSe) & E(logitSp)    .0045838
```

The bivariate and HSROC parameter estimates are displayed along with their standard errors and approximate 95% confidence intervals in the standard Stata format. The bivariate location parameters, μ_A and μ_B, are denoted by `E(logitSe)` and `E(logitSp)`; the variance parameters, σ_A^2 and σ_B^2, are shown as `Var(logitSe)` and `Var(logitSp)`; and the correlation, σ_{AB}, is shown as `Corr(logits)`. The HSROC parameters are denoted by using the notation of Rutter and Gatsonis (2001) given in section 3.1, spelling out Greek letters with capital initials for the capital Greek letters Λ and Θ, and showing σ_α^2 and σ_θ^2 as `s2alpha` and `s2theta`.

z statistics and p-values are not given for most of the parameters because parameter values of zero do not correspond to null hypotheses of interest. The exception is the HSROC shape (asymmetry) parameter, β (beta), where $\beta = 0$ corresponds to a symmetric ROC curve in which the diagnostic odds ratio does not vary along the curve.

The output also gives summary values and confidence intervals for the sensitivity (Se) and specificity (Sp) (back-transformed from E(logitSe) and E(logitSp)), as well as values for the diagnostic odds ratio (DOR) and the positive and negative likelihood ratios (LR+ and LR-) at the summary point. The summary likelihood ratios will not, in general, be the same as would be obtained by first calculating the likelihood ratios for each study and meta-analyzing these. Such an approach has been deprecated in favor of the approach implemented here (Zwinderman and Bossuyt 2008). A summary value for the inverse of the negative likelihood ratio (1/LR-) is also given, because larger values of the inverse of the negative likelihood ratio indicate a more accurate test, and comparing this with the positive likelihood ratio can indicate whether a positive or negative test result has greater impact on the odds of disease.

Finally, the output shows the covariance between $\widehat{\mu}_A$ and $\widehat{\mu}_B$. This is needed to draw confidence and prediction regions, and is included to make it easier to do so in external software, such as the Cochrane Collaboration's Review Manager 5 (Nordic Cochrane Centre 2007).

❏ **Technical note**

On rare occasions, during model fitting, gllamm may report an error, such as "convergence not achieved: try with more quadrature points" or (less transparently) "log likelihood cannot be computed". Increasing the number of integration points beyond metandi's default of 5 by using the nip() option (e.g., nip(7)) may resolve this.

❏

5   metandiplot

The metandiplot command produces a graph of the model fit by metandi, which must be the last estimation-class command executed. For convenience, the metandi command has a plot option, which produces the same graph. If metandiplot is not followed by a varlist, then the study-specific estimates (shown by the circles in figure 2) are not included in the graph. The metandiplot command has options to alter the default appearance of the graph or to turn off any of the plot elements. These options are not available when using the plot option to metandi. metandiplot can be run many times with different options without refitting the model with metandi.

```
. metandiplot tp fp fn tn
```

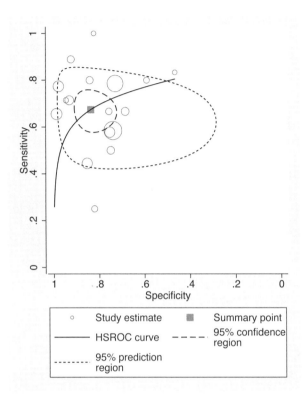

Figure 2. Plot of fitted model from `metandiplot`

The resulting graph (figure 2) shows the following summaries, together with circles showing the individual study estimates:

- A summary curve from the HSROC model

- A summary operating point, i.e., summary values for sensitivity and specificity

- A 95% confidence region for the summary operating point

- A 95% prediction region (confidence region for a forecast of the true sensitivity and specificity in a future study)

The default is to include all the summaries listed above, which can result in a rather cluttered graph, so options are included to remove any of the elements; for example, `predopts(off)` turns off the prediction region. See section 7.2 for more information about `metandiplot` options.

By default, the summary HSROC curve is displayed only for sensitivities and specificities at least as large as the smallest study-specific estimates if a varlist is included.

The shape of the prediction region is dependent on the assumption of a bivariate normal distribution for the random effects and should therefore not be overinterpreted; it is intended to give a visual representation of the extent of between-study heterogeneity, which is often considerable.

6   predict after metandi

Many of Stata's standard postestimation tools will not work after `metandi` or will not work as expected, because `metandi` temporarily reshapes the data before fitting the model.

The notable exception is `predict`, which can be used to obtain posterior predictions (empirical Bayes estimates) of the sensitivity and specificity in each study (`mu`), as well as various statistics that can be useful for detecting outliers (e.g., `ustd`) and influential observations (`cooksd`).

The help file provides basic commands for examining diagnostics. We take the opportunity here to provide slightly more customized displays.

Empirical Bayes estimates give the best estimate of the true sensitivity and specificity in each study, and these estimates will be "shrunk" toward the summary point compared with the study-specific estimates shown in figure 1.

```
. predict eb
(option mu assumed; posterior predicted Se & Sp)

. metandiplot, addplot(scatter eb1 eb0, msymbol(o))
> legend(label(5 "Empirical Bayes"))
```

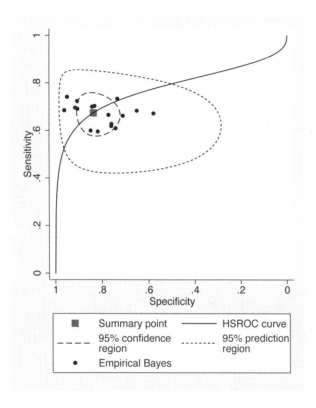

Figure 3. Empirical Bayes estimates

Comparing figure 3 with figure 2 shows that the shrinkage is generally greater for sensitivity than for specificity in this example, reflecting both the smaller variance of sensitivity (on the logit scale) and the fact that most studies have fewer participants with disease than without disease, leading to more precise estimates of specificity than of sensitivity.

Cook's distance is a measure of the influence of a study on the model parameters and can be used to check for particularly influential studies. Cook's distance is calculated using `gllapred` and so is available in Stata 10 only if the `gllamm` option was used with `metandi`. `gllapred` calculates Cook's distance to measure influence on all model parameters including the variance parameters (Skrondal and Rabe-Hesketh 2004, sec. 8.6.6). To check for outliers, standardized predicted random effects can be interpreted as standardized study-level residuals.

```
. metandi tp fp fn tn, gllamm nolog
  (output omitted)
. predict cooksd, cooksd
(Cook's distance may take a few seconds...)
. predict ustd_Se ustd_Sp, ustd
. local opts "mlabel(studyid) mlabpos(0) m(i) nodraw"
. scatter cooksd studyid,'opts' name(cooksd)
. scatter ustd_Se ustd_Sp, xscale(rev) xla(, grid) xline(0) yline(0) 'opts'
> name(ustd)
. graph combine cooksd ustd, xsize(5) scale(*1.5)
```

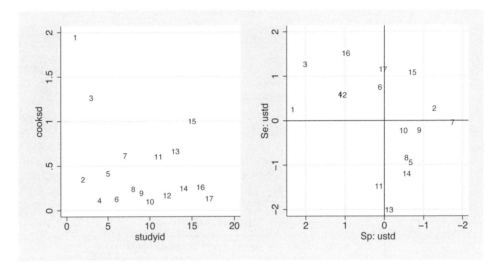

Figure 4. Left panel: Cook's distance. Right panel: Standardized residuals (standardized predicted random effects).

Figure 4 shows both Cook's distance and the standardized residuals. (The residual corresponding to specificity has been plotted on a reversed axis to correspond with the convention for ROC plots used in figure 1.) These two graphs are best read in combination. Cook's distance shows which studies are influential, while the standardized residuals give some insight into why. According to Skrondal and Rabe-Hesketh (2004), a typical cutpoint for declaring a value of Cook's D to be "large" is four times the number of parameters divided by the number of clusters (here studies). (Definitions of Cook's D differ, hence so does the cutpoint—the definition used by Stata in [R] **regress postestimation** divides by the number of parameters.) Because there are five parameters in this model, this suggests a cutpoint of 20 divided by the number of studies for interpreting Cook's D after `metandi`, giving $20/17 \approx 1.2$ for the lymphangiography meta-analysis. Here, study 1 is particularly influential, followed by study 3. Studies 1 and 3 have high standardized residuals for specificity, leading to influence on both the mean and variance of logit-transformed specificity. Study 13 has a large (negative) standardized residual for sensitivity but does not appear to be so influential as judged by its Cook's distance. Further investigation of the effect of individual studies on the model could be undertaken by refitting the model and leaving out each study in turn.

7   Syntax and options for commands

7.1   The metandi command

Syntax

metandi *tp fp fn tn* $\begin{bmatrix} if \end{bmatrix}$ $\begin{bmatrix} in \end{bmatrix}$ $\begin{bmatrix} , \underline{p}lot \underline{g}llamm force ip(g|m) \underline{nip}(\#) \end{bmatrix}$

    <u>nobi</u>variate <u>noh</u>sroc <u>nos</u>ummarypt <u>d</u>etail <u>l</u>evel(\#) <u>trace</u> <u>nolog</u>$\Big]$

    by is allowed with metandi; see [D] **by**.

Options

plot requests a plot of the results on an SROC plot. This is a convenience option equivalent to executing the metandiplot command after metandi with the same list of variables, *tp*, *fp*, *fn*, and *tn* (and the same *if* and *in* qualifiers, if specified). Greater control of the plot is available through the options of the metandiplot command when issued as a separate command after metandi.

gllamm specifies that the model be fit using gllamm. This is the default in Stata 8 and 9, so the option is of use only in Stata 10, in which the model is fit using xtmelogit by default.

force forces metandi to attempt to fit data where one or more studies have $tp + fn = 0$ (or $tn + fp = 0$), i.e., where there are no individuals that are positive (negative) for the reference standard. Without this option, metandi exits with an error when such data exist. Problems may be encountered in fitting such data, particularly when the model is fit using xtmelogit. Sensitivity (specificity) cannot be estimated within such studies, so they are not included in the plot produced by metandiplot.

ip(g|m) specifies the quadrature (numerical integration) method used to integrate out the random effects: ip(g), the default, gives Cartesian product quadrature, while ip(m) gives spherical quadrature, which is available in gllamm but not in xtmelogit. Spherical quadrature can be more efficient, though its properties are less well known and it can sometimes cause the adaptive quadrature step to take longer to converge. See Rabe-Hesketh, Skrondal, and Pickles (2005).

nip(\#) specifies the number of integration points used for quadrature. Higher values should result in greater accuracy but typically at the expense of longer execution times. Specifying too small a value can lead to convergence problems or even failure of adaptive quadrature; if you receive the error "log likelihood cannot be computed", try increasing nip(). For Cartesian product quadrature, nip() specifies the number of points for each of the two random effects; the default is nip(5). For spher-

ical quadrature, `nip()` specifies the degree, d, of the approximation; the default is
`nip(9)`, and the only values currently supported by `gllamm` are 5, 7, 9, 11, and 15.
These defaults give approximately the same accuracy, because degree d for spherical
quadrature approximately corresponds in accuracy to $(d+1)/2$ points per random
effect for Cartesian product quadrature (Rabe-Hesketh, Skrondal, and Pickles 2005,
app. B).

`nobivariate`, `nohsroc`, and `nosummarypt` suppress reporting of the bivariate parameter estimates, the HSROC parameter estimates, or the summary point estimates,
respectively.

`detail` displays the output of all `gllamm` or `xtmelogit` commands issued.

`level(#)` specifies the confidence level, as a percentage, for confidence intervals. The
default is `level(95)` or as set by `set level`; see [U] **20.7 Specifying the width
of confidence intervals**.

`trace` adds a display of the current parameter vector to the iteration log.

`nolog` suppresses display of the iteration log.

7.2   The metandiplot command

Syntax

`metandiplot` [*tp fp fn tn*] [*if*] [*in*] [*weight*] [, not̲runcate le̲vel(*#*)
    p̲redlevel(*numlist*) n̲points(*#*) *subplot_options* addplot(*plot*)
    twoway_options]

Options

not̲runcate specifies that the HSROC curve will not be truncated outside the region
of the data. By default, the HSROC curve is not shown when the sensitivity or
specificity is less than its smallest study estimate.

le̲vel(*#*) specifies the confidence level, as a percentage, for the confidence contour.
The default is `level(95)` or as set by `set level`; see [U] **20.7 Specifying the
width of confidence intervals**.

p̲redlevel(*numlist*) specifies the levels, as a percentage, for the prediction contour(s).
The default is one contour at the same probability level as the confidence region.
Up to five prediction contours are allowed.

n̲points(*#*) specifies the number of points to use in drawing the outlines of the confidence and prediction regions. The default is `npoints(500)`.

subplot_options, which are s̲ummopts(), c̲onfopts(), p̲redopts(), c̲urveopts(), and
s̲tudyopts(), control the display of the summary point, confidence contour, prediction contour(s), HSROC curve, and study symbols, respectively. The options within

each set of parentheses are simply passed through to the appropriate `twoway` plot. Any of the plots can be turned off by specifying, for example, `summopts(off)`.

`addplot(`*plot*`)` allows adding additional `graph twoway` plots to the graph; see [G] *add-plot_option*. For example, empirical Bayes predictions could be generated by using `predict` after `metandi` and then added to the graph. See section 6.

twoway_options are most of the options documented in [G] ***twoway_options***, including options for titles, axes, labels, schemes, and saving the graph to disk. However, the `by()` option is not allowed.

7.3 The predict command after metandi

Syntax

`predict` [*type*] *newvarlist* [*if*] [*in*] [, *statistic*]

statistic	description
`mu`	posterior predicted (empirical Bayes) sensitivity and specificity; the default
`u`	posterior means (empirical Bayes predictions, BLUPs) of random effects
`sdu`	posterior standard deviations of random effects
`ustd`	standardized posterior means of random effects
`linpred`	linear predictor with empirical Bayes predictions plugged in: `linpred = xb + u`
`cooksd`	Cook's distance for each study; available only when model was fit using `gllamm`

Most of the above statistics require *newvarlist* to consist of two new variables to store them: one for the statistic associated with sensitivity and one for the statistic associated with specificity. If *newvarlist* contains only one *newvar*, the statistics associated with sensitivity and specificity will be stored in *newvar*1 and *newvar*0, respectively. `cooksd`, however, is computed once for each study and therefore requires only one *newvar*. See section 6 for examples.

(Continued on next page)

7.4   Saved results

`metandi` saves the following results in `e()`:

Scalars
`e(N)`	number of studies	`e(ll)`	log likelihood

Macros
`e(cmd)`	`metandi`	`e(predict)`	program used to implement
`e(tpfpfntn)`	names of *tp fp fn tn*		predict
	variables	`e(properties)` b V	
`e(cmd)`	`metareg`		

Matrices
`e(b)`	bivariate coefficient vector	`e(V)`	variance–covariance matrix of
`e(b_hsroc)`	HSROC coefficient vector		the bivariate estimators
		`e(V_hsroc)`	variance–covariance matrix of
			the HSROC estimators

Functions
`e(sample)`	marks estimation sample

8   Methods and formulas

It is possible to use routines for linear mixed models to fit an approximate version of the bivariate model obtained by using empirical logit transforms of the estimated sensitivity and specificity in each study together with their estimated standard errors (Reitsma et al. 2005). However, the small cell counts common in diagnostic accuracy studies can lead to poor performance of such approximations. Generalized mixed models, in particular, hierarchical (mixed-effects) logistic regression, can handle the binomial nature of the data directly and are therefore preferable (Chu and Cole 2006; Riley et al. 2007).

Such models are complex to fit, however, because they require numerical integration (quadrature) to integrate out the random effects. `metandi` uses `gllamm` or `xtmelogit` to fit the bivariate model by using adaptive quadrature, then transforms the parameter estimates to those of the HSROC model by using the delta method (Cox 1998).

Because the bivariate model can sometimes prove difficult to fit, some care has been taken to provide good starting values. First, two separate univariate models are fit to sensitivity and specificity. These provide excellent starting values for the two mean and two variance parameters of the bivariate model. A reasonable starting value for the correlation parameter is obtained from the correlation between the posterior means (empirical Bayes predictions) of the two univariate models.

We now give the mathematical forms of the bivariate and HSROC models in the absence of covariates. See Rutter and Gatsonis (2001); Reitsma et al. (2005); and Harbord et al. (2007) for information on the models with covariates, which are not currently supported by `metandi`.

8.1   The bivariate model

Following Reitsma et al. (2005), we denote the sensitivity in the ith study by p_{Ai} and the specificity by p_{Bi}, and base analysis on their logit transforms:

$$\mu_{Ai} = \text{logit}(p_{Ai})$$

$$\mu_{Bi} = \text{logit}(p_{Bi})$$

(We use the letter μ where Reitsma et al. (2005) used θ to avoid a clash of notation with the HSROC model defined in the next section.)

The bivariate model is a random-effects model in which the logit transforms of the true sensitivity and true specificity in each study have a bivariate normal distribution across studies, thereby allowing for the possibility of correlation between them (Reitsma et al. 2005):

$$\begin{pmatrix} \mu_{Ai} \\ \mu_{Bi} \end{pmatrix} \sim N \left\{ \begin{pmatrix} \mu_A \\ \mu_B \end{pmatrix}, \Sigma_{AB} \right\} \quad \text{with} \quad \Sigma_{AB} = \begin{pmatrix} \sigma_A^2 & \sigma_{AB} \\ \sigma_{AB} & \sigma_B^2 \end{pmatrix}$$

8.2   The HSROC model

The HSROC model (Rutter and Gatsonis 2001) was originally formulated in terms of the probability, π_{ij}, that a patient in study i with disease status j has a positive test result, where $j = 0$ for a patient without the disease and $j = 1$ for a patient with the disease. Therefore, sensitivity $p_{Ai} = \pi_{i1}$ and specificity $p_{Bi} = 1 - \pi_{i0}$.

The HSROC model for study i takes the form

$$\text{logit}(\pi_{ij}) = (\theta_i + \alpha_i X_{ij}) \exp(-\beta X_{ij}) \tag{1}$$

where $X_{ij} = -1/2$ for those without disease ($j = 0$) and $+1/2$ for those with disease ($j = 1$). Both θ_i and α_i are allowed to vary between studies. In the model without covariates fit by `metandi`, they are assumed to have independent normal distributions with $\theta_i \sim N(\Theta, \sigma_\theta^2)$ and $\alpha_i \sim N(\Lambda, \sigma_\alpha^2)$. The model is nonlinear in the parameter β and therefore cannot be fit in `gllamm` directly.

We can rewrite (1) as two separate equations for the logit transforms of sensitivity p_{Ai} and specificity p_{Bi}, thus connecting to the parameters μ_{Ai} and μ_{Bi} of the bivariate model above:

$$\mu_{Ai} = \text{logit}(p_{Ai}) = b^{-1}(\theta_i + \frac{1}{2}\alpha_i)$$

$$\mu_{Bi} = \text{logit}(p_{Bi}) = -b(\theta_i - \frac{1}{2}\alpha_i)$$

This tells us that μ_{Ai} and μ_{Bi} are linear combinations of two random variables, θ_i and α_i, with independent normal distributions, and that they therefore must have a bivariate normal distribution. Some straightforward further algebra gives the explicit relationship between the parameters of the two models (Harbord et al. 2007; Arends et al.

2008), enabling HSROC parameter estimates to be obtained by transforming the bivariate parameter estimates. Standard errors for the transformed parameter estimates are obtained by the delta method, which gives the same standard errors that would be obtained from standard maximum-likelihood methods if the HSROC model were fit directly (Cox 1998).

8.3   Methods and formulas for metandiplot

HSROC curve

The HSROC model gives rise to an SROC curve by allowing the threshold parameter, θ_i, to vary while holding the accuracy parameter, α_i, fixed at its mean, Λ. The expected sensitivity for a given specificity is then given by (Rutter and Gatsonis 2001; Macaskill 2004)

$$\text{logit(sensitivity)} = \Lambda e^{-\beta/2} - e^{-\beta}\text{logit(specificity)}$$

Bivariate confidence and prediction regions

Confidence and prediction regions in SROC space can be constructed by using the estimates from the bivariate model (Reitsma et al. 2005; Harbord et al. 2007). An elliptical joint confidence region for μ_A and μ_B is most easily specified by using a parametric representation (Douglas 1993)

$$\mu_A = \widehat{\mu}_A + s_A\, c\, \cos t \tag{2}$$

$$\mu_B = \widehat{\mu}_B + s_B\, c\, \cos(t + \arccos r) \tag{3}$$

where s_A and s_B are the estimated standard errors of $\widehat{\mu}_A$ and $\widehat{\mu}_B$, r is the estimate of their correlation, and varying t from 0 to 2π generates the boundary of the ellipse. The constant c has been called the boundary constant of the ellipse (Alexandersson 2004); $c = \sqrt{2f_{2,n-2;\alpha}}$, where n is the number of studies and $f_{2,n-2;\alpha}$ is the upper $100\alpha\%$ point of the F distribution with degrees of freedom 2 and $n-2$ (Douglas 1993; Chew 1966). This ellipse is then back-transformed to conventional ROC space to give a confidence region for the summary operating point.

A prediction region giving the region that has a given probability (e.g., 95%) of including the *true* sensitivity and specificity of a future study is generated similarly. The covariance matrix for the true logit sensitivity and logit specificity in a future study is

$$\Sigma_{AB} + \text{Var}\begin{pmatrix}\widehat{\mu}_A \\ \widehat{\mu}_B\end{pmatrix}$$

In practice, both terms are estimated by fitting the model to the data. The parameters s_A, s_B, and r in (2) and (3) can then be replaced by the corresponding quantities derived from this covariance matrix to give the prediction ellipse in logit ROC space, which is then back-transformed to a prediction region for the true sensitivity and specificity of a future study in conventional ROC space.

8.4 Methods and formulas for predict

If `metandi` fit the model by using `gllamm`, then `predict` after `metandi` uses `gllapred`; see Rabe-Hesketh, Skrondal, and Pickles (2004). If `metandi` fit the model by using `xtmelogit`, `predict` after `metandi` uses the prediction facilities of `xtmelogit`; see [XT] **xtmelogit postestimation**.

9 Acknowledgments

Joseph Coveney first worked out how to fit the bivariate model by using `gllamm` and posted the syntax on Statalist in response to a query from Ben Dwamena; our thanks to Joe for generous email correspondence. We thank the authors of `gllamm` for all their work, and Sophia Rabe-Hesketh in particular for helpful email correspondence. Our thanks also to Susan Mallett and Jon Deeks for useful feedback on earlier versions of `metandi`.

10 References

Alexandersson, A. 2004. Graphing confidence ellipses: An update of ellip for Stata 8. *Stata Journal* 4: 242–256.

Arends, L. R., T. H. Hamza, J. C. van Houwelingen, M. H. Heijenbrok-Kal, M. G. M. Hunink, and T. Stijnen. 2008. Bivariate random effects meta-analysis of ROC curves. *Medical Decision Making* 28: 621–638.

Bradburn, M. J., J. J. Deeks, and D. G. Altman. 1998. sbe24: metan—an alternative meta-analysis command. *Stata Technical Bulletin* 44: 4–15. Reprinted in *Stata Technical Bulletin Reprints*, vol. 8, pp. 86–100. College Station, TX: Stata Press. (Updated article is reprinted in this collection on pp. 3–28.)

Chew, V. 1966. Confidence, prediction, and tolerance regions for the multivariate normal distribution. *Journal of the American Statistical Association* 61: 605–617.

Chu, H., and S. R. Cole. 2006. Bivariate meta-analysis of sensitivity and specificity with sparse data: A generalized linear mixed model approach. *Journal of Clinical Epidemiology* 59: 1331–1332.

Coveney, J. 2004. Re: st: bivariate random effects meta-analysis of diagnostic test. Statalist archive. Available at http://www.stata.com/statalist/archive/2004-04/msg00820.html.

Cox, C. 1998. Delta method. In *Encyclopedia of Biostatistics*, ed. P. Armitage and T. Colton, 1125–1127. New York: Wiley.

Deeks, J. J. 2001a. Systematic reviews of evaluations of diagnostic and screening tests. In *Systematic Reviews in Health Care: Meta-Analysis in Context*, 2nd edition, ed. M. Egger, G. Davey Smith, and D. G. Altman, 248–282. London: BMJ Books.

———. 2001b. Systematic reviews in health care: Systematic reviews of evaluations of diagnostic and screening tests. *British Medical Journal* 323: 157–162.

Devillé, W. L., F. Buntinx, L. M. Bouter, V. M. Montori, H. C. W. de Vet, D. A. W. M. van der Windt, and P. D. Bezemer. 2002. Conducting systematic reviews of diagnostic studies: Didactic guidelines. *BMC Medical Research Methodology* 2: 9.

Douglas, J. B. 1993. Confidence regions for parameter pairs. *American Statistician* 47: 43–45.

Gatsonis, C., and P. Paliwal. 2006. Meta-analysis of diagnostic and screening test accuracy evaluations: Methodologic primer. *American Journal of Roentgenology* 187: 271–281.

Gluud, C., and L. L. Gluud. 2005. Evidence based diagnostics. *British Medical Journal* 330: 724–726.

Harbord, R. M., J. J. Deeks, M. Egger, P. Whiting, and J. A. C. Sterne. 2007. A unification of models for meta-analysis of diagnostic accuracy studies. *Biostatistics* 8: 239–251.

Lijmer, J. G., P. M. M. Bossuyt, and S. H. Heisterkamp. 2002. Exploring sources of heterogeneity in systematic reviews of diagnostic tests. *Statistics in Medicine* 21: 1525–1537.

Macaskill, P. 2004. Empirical Bayes estimates generated in a hierarchical summary ROC analysis agreed closely with those of a full Bayesian analysis. *Journal of Clinical Epidemiology* 57: 925–932.

Mallett, S., J. J. Deeks, S. Halligan, S. Hopewell, V. Cornelius, and D. G. Altman. 2006. Systematic reviews of diagnostic tests in cancer: Review of methods and reporting. *British Medical Journal* 333: 413–416.

Moses, L. E., D. Shapiro, and B. Littenberg. 1993. Combining independent studies of a diagnostic test into a summary ROC curve: Data-analytic approaches and some additional considerations. *Statistics in Medicine* 12: 1293–1316.

Nordic Cochrane Centre. 2007. *Review Manager (RevMan): Version 5.* Software program. Copenhagen: The Nordic Cochrane Centre, The Cochrane Collaboration.

Rabe-Hesketh, S., A. Skrondal, and A. Pickles. 2004. GLLAMM manual. Working Paper 160, Division of Biostatistics, University of California–Berkeley. Downloadable from http://www.bepress.com/ucbbiostat/paper160/.

———. 2005. Maximum likelihood estimation of limited and discrete dependent variable models with nested random effects. *Journal of Econometrics* 128: 301–323.

Reitsma, J. B., A. S. Glas, A. W. S. Rutjes, R. J. P. M. Scholten, P. M. Bossuyt, and A. H. Zwinderman. 2005. Bivariate analysis of sensitivity and specificity produces informative summary measures in diagnostic reviews. *Journal of Clinical Epidemiology* 58: 982–990.

Riley, R. D., K. R. Abrams, A. J. Sutton, P. C. Lambert, and J. R. Thompson. 2007. Bivariate random-effects meta-analysis and the estimation of between-study correlation. *BMC Medical Research Methodology* 7: 3.

Rutter, C. M., and C. A. Gatsonis. 2001. A hierarchical regression approach to meta-analysis of diagnostic test accuracy evaluations. *Statistics in Medicine* 20: 2865–2884.

Scheidler, J., H. Hricak, K. K. Yu, L. Subak, and M. R. Segal. 1997. Radiological evaluation of lymph node metastases in patients with cervical cancer: A meta-analysis. *Journal of the American Medical Association* 278: 1096–1101.

Skrondal, A., and S. Rabe-Hesketh. 2004. *Generalized Latent Variable Modeling: Multilevel, Longitudinal, and Structural Equation Models.* Boca Raton, FL: Chapman & Hall/CRC.

Sterne, J. A. C., R. J. Harris, R. M. Harbord, and T. J. Steichen. 2007. What meta-analysis features are available in Stata? Stata FAQ. Available at http://www.stata.com/support/faqs/stat/meta.html.

Tatsioni, A., D. A. Zarin, N. Aronson, D. J. Samson, C. R. Flamm, C. Schmid, and J. Lau. 2005. Challenges in systematic reviews of diagnostic technologies. *Annals of Internal Medicine* 142: 1048–1055.

Walter, S. D. 2002. Properties of the summary receiver operating characteric (SROC) curve for diagnostic test data. *Statistics in Medicine* 21: 1237–1256.

Zwinderman, A. H., and P. M. Bossuyt. 2008. We should not pool diagnostic likelihood ratios in systematic reviews. *Statistics in Medicine* 27: 687–697.

The Stata Journal (2006)
6, Number 1, pp. 40–57

Generalized least squares for trend estimation of summarized dose–response data

Nicola Orsini
Karolinska Institutet
Stockholm, Sweden
nicola.orsini@ki.se

Rino Bellocco
Karolinska Institutet
Stockholm, Sweden
and Department of Statistics
University of Milano-Bicocca
rino.bellocco@ki.se

Sander Greenland
Department of Epidemiology and Department of Statistics
University of California
Los Angeles, CA
lesdomes@ucla.edu

Abstract. This paper presents a command, glst, for trend estimation across different exposure levels for either single or multiple summarized case–control, incidence-rate, and cumulative incidence data. This approach is based on constructing an approximate covariance estimate for the log relative risks and estimating a corrected linear trend using generalized least squares. For trend analysis of multiple studies, glst can estimate fixed- and random-effects meta-regression models.

Keywords: st0096, glst, dose–response data, generalized least squares, trend, meta-analysis, meta-regression

1 Introduction

Epidemiological studies often assess whether the observed relationship between increasing (or decreasing) levels of exposure and the risk (or odds) of diseases follows a linear dose–response pattern. Methods for trend estimation of single and multiple summarized dose–response studies (Berlin, Longnecker, and Greenland 1993) are particularly useful when the full original data are not available.

To demonstrate these methods, our paper uses different types of dose–response data arising from published case–control, incidence-rate, and cumulative incidence data (also see [ST] **epitab**). Summarized data are typically reported as a series of dose-specific relative risks, with one category serving as the common referent group. The term *relative risk* (RR) will be used as a generic term for the risk ratio (cumulative incidence data), rate ratio (incidence-rate data), and odds ratio (case–control data).

Table 1 shows a summary of case–control data investigating the association between the consumption of alcohol and the risk of breast cancer, first presented by

Rohan and McMichael (1988), in which it appears that risk of breast cancer increases with increasing levels of alcohol intake.

Table 1. Case–control data on alcohol and breast cancer risk (Rohan and McMichael 1988)

Alcohol (g/d)	Assigned dose (g/d)	No. of cases	No. of controls	Total subjects	Adjusted RR (95% CI)
0	0	165	172	337	1.0 (Referent)
<2.5	2	74	93	167	0.80 (0.51–1.27)
2.5–9.3	6	90	96	186	1.16 (0.73–1.85)
>9.3	11	122	90	212	1.57 (0.99–2.51)

Table 2 shows a summary of incidence-rate data investigating the association between the long-term intake of dietary fiber and risk of coronary heart disease among women, first presented by Wolk et al. (1999), which supports the hypothesis that higher fiber intake

Table 2. Incidence-rate data on dietary fiber and coronary heart disease risk (Wolk et al. 1999)

	Adjusted RR (95% CI)
	1.0 (Referent)
	0.98 (0.77–1.24)
	0.92 (0.71–1.18)
	0.87 (0.66–1.15)
	0.77 (0.57–1.04)

Table ... investigating the association between ... al cancer, first presented by Larsson, ... more servings per day of high-fat dairy ...

Table 3. Cumulative incidence data on high-fat dairy food and colorectal cancer risk (Larsson, Bergkvist, and Wolk 2005)

High-fat dairy (servings/d)	Assigned dose (servings/d)	No. of cases	Total subjects	Adjusted RR (95% CI)
< 1.0	0.5	110	8,103	1.0 (Referent)
1.0– <2.0	1.5	212	17,538	0.75 (0.60–0.96)
2.0– <3.0	2.5	211	15,304	0.74 (0.58–0.95)
3.0– <4.0	3.5	132	9,078	0.68 (0.52–0.90)
≥4.0	6.5	133	10,685	0.59 (0.44–0.79)

For each of these summarized tables, we have adjusted relative risks and confidence limits for each nonreference exposure level. The usual approach to trend estimation, namely, the expected change of the log relative risks for a unit change of the exposure level, is to fit a linear regression through the origin, where the response variable is the log relative risks, the assigned dose is the covariate, and the log relative risks are weighted by the inverse of their variances. This method is known as weighted least-squares (WLS) regression (see [R] **vwls**), and it assumes that the log relative risks are independent—an assumption that is never satisfied in practice. The log relative risks are correlated given that they are estimated using a common referent group, and this standard approach underestimates the variance of the slope (Greenland and Longnecker 1992). This problem can be particularly relevant in a meta-analysis of summarized dose–response data where each study slope (trend) is weighted by the inverse of the variance (Shi and Copas 2004).

An efficient estimation method for the slope of a single study is therefore proposed and implemented in the command glst, as described by Greenland and Longnecker (1992). This method is then incorporated in the estimation of fixed and random-effects meta-regression models for the analysis of multiple studies.

The rest of the article is organized as follows: section 2 introduces the dose–response model and the estimation method; section 3 describes the syntax of the command glst; section 4 presents some practical examples based on published data; section 5 compares the corrected and uncorrected methods for trend estimation; and section 6 contains final comments.

2   Method

2.1   Log-linear dose–response model for a single study

It is possible to analyze the shape of the dose–response relationship between reported log relative risks and the exposure levels by estimating a log-linear dose–response regression model (Greenland and Longnecker 1992; Berlin, Longnecker, and Greenland 1993;

Shi and Copas 2004). Assuming that the exposure variable takes value 0 in the reference category, the estimated log relative risk in the reference category is set to zero (log 1); therefore, no intercept models are used. The matrix notation is

$$\mathbf{y} = \mathbf{X}\beta + \mathbf{e} \tag{1}$$

$$\mathbf{y} = \begin{bmatrix} y_1 \\ \vdots \\ y_i \\ \vdots \\ y_n \end{bmatrix} \quad \mathbf{X} = \begin{bmatrix} x_{11} & x_{12} & \cdots & x_{1p} \\ \vdots & \vdots & & \vdots \\ x_{i1} & x_{i2} & & x_{ip} \\ \vdots & \vdots & & \vdots \\ x_{n1} & x_{n2} & \cdots & x_{np} \end{bmatrix} \quad \beta = \begin{bmatrix} \beta_1 \\ \vdots \\ \beta_p \end{bmatrix} \quad \mathbf{e} = \begin{bmatrix} \varepsilon_1 \\ \vdots \\ \varepsilon_i \\ \vdots \\ \varepsilon_n \end{bmatrix}$$

where $\mathbf{y}$ is an $n \times 1$ vector of (reported) estimated log relative risks; $i = 1, 2, \ldots, n$ identifies nonreference exposure levels; $\mathbf{X}$ is an $n \times p$ matrix of nonstochastic covariates, where the first column, denoted by x_{i1}, identifies the exposure variable, and the remaining $p - 1$ columns, for instance, may represent transformations of x_{i1}; β is a $p \times 1$ vector of unknown regression coefficients; and $\mathbf{e}$ is an $n \times 1$ vector of random errors, with expected value $E(\mathbf{e}) = 0$ and variance–covariance matrix $\mathrm{Cov}(\mathbf{e}) = E(\mathbf{e}\mathbf{e}')$ equal to the following symmetric matrix given by

$$\mathrm{Cov}(\mathbf{e}) = \mathbf{\Sigma} = \begin{bmatrix} \sigma_{11} & & & & \\ \vdots & \ddots & & & \\ \sigma_{i1} & & \sigma_{ij} & & \\ \vdots & & & \ddots & \\ \sigma_{n1} & \cdots & \sigma_{nj} & \cdots & \sigma_{nn} \end{bmatrix}$$

Thus the response variable $\mathbf{y}$ has expected value $E(\mathbf{y}) = \mathbf{X}\beta$ and covariance matrix $\mathrm{Cov}(\mathbf{y}) = \mathbf{\Sigma}$.

2.2 Generalized least squares

We use generalized least squares (GLS) to efficiently estimate the β vector of regression coefficients in (1). Assuming that the variance–covariance matrix of $\mathbf{e}$ is $\mathrm{Cov}(\mathbf{e}) = \mathbf{\Sigma}$, this method involves minimizing $(\mathbf{y} - \mathbf{X}\beta)'\mathbf{\Sigma}^{-1}(\mathbf{y} - \mathbf{X}\beta)$ with respect to β. Suppose initially that the variance–covariance matrix $\mathbf{\Sigma}$ is known. In matrix notation, the resulting estimator $\mathbf{b}$ of the regression coefficients β is

$$\mathbf{b} = (\mathbf{X}'\mathbf{\Sigma}^{-1}\mathbf{X})^{-1}\mathbf{X}'\mathbf{\Sigma}^{-1}\mathbf{y} \tag{2}$$

and the estimated covariance matrix $\mathbf{v}$ of $\mathbf{b}$ is

$$\mathbf{v} = \widehat{\mathrm{Cov}}(\mathbf{b}) = (\mathbf{X}'\mathbf{\Sigma}^{-1}\mathbf{X})^{-1} \tag{3}$$

A remarkable property of the GLS estimator is that for any choice of $\mathbf{\Sigma}$, the GLS estimate of β is unbiased; that is, $E(\mathbf{b}) = \beta$.

GLS estimation imposes no distributional assumption for the random errors, $\mathbf{e}$, whereas maximum likelihood (ML) estimation assumes a distribution, and the log likelihood of the sample observed is then maximized. Under the assumption that random errors are normally distributed with zero mean and variance–covariance matrix $\mathbf{\Sigma}$, i.e., $\mathbf{e} \sim N(0, \mathbf{\Sigma})$, the log-likelihood function can be written as the following:

$$l = -\frac{n}{2} \log(2\pi) - \frac{1}{2} \log |\mathbf{\Sigma}| - \frac{1}{2} \left\{ (\mathbf{y} - \mathbf{X}\beta)' \mathbf{\Sigma}^{-1} (\mathbf{y} - \mathbf{X}\beta) \right\} \tag{4}$$

Maximizing (4) with respect to β is equivalent to solving $\partial l / \partial \beta = 0$. The solution is the ML estimator of β, which under the normality assumption turns out to be the same as the GLS estimator given by (2).

2.3 Statistical inference

To construct confidence intervals and tests of hypotheses about β, we can make direct use of the GLS estimate, $\mathbf{b}$, and its estimated covariance matrix, $\mathbf{v}$. When the normality assumption of the random error $\mathbf{e}$ is introduced, the distributional properties of $\mathbf{y}$ and functions of $\mathbf{y}$ follow at once.

Because $\mathbf{y} \sim N(\mathbf{X}\beta, \mathbf{\Sigma})$, the vector $\mathbf{b}$, which is a linear function of $\mathbf{y}$, is therefore approximately normally distributed $\mathbf{b} \sim N(\beta, \mathbf{v})$.

A test of the null hypothesis, H_0: $\mathbf{b}_j = 0$ versus H_A: $\mathbf{b}_j \neq 0$, can be based on the following Wald statistic,

$$Z = \frac{\mathbf{b}_j}{\sqrt{\mathbf{v}_j}}$$

where $\mathbf{b}_j$ denotes the jth element of the vector $\mathbf{b}$ and $\mathbf{v}_j$ denotes the jth diagonal element of $\mathbf{v}$, with $j = 1, 2, \ldots, p$. The Z statistic can be compared with a standard normal distribution.

Wald test–type confidence intervals of β are computed using the large-sample approximation, the z distribution rather than the t distribution, because the estimates, $\mathbf{b}$, are based on a collection of n presumably large groups of subjects rather than n subjects (Grizzle, Starmer, and Koch 1969; Greenland 1987).

2.4 Covariances

In summarized dose–response data, the log relative risks, $\mathbf{y}$, are estimated using a common reference group. Therefore, the elements of $\mathbf{y}$ are not independent and the off-diagonal elements of $\mathbf{\Sigma}$ are not zero (Greenland and Longnecker 1992). This section describes the method and formulas needed to estimate all the elements of $\mathbf{\Sigma}$.

The diagonal element σ_{ii} of $\mathbf{\Sigma}$, the variance of the log relative risk y_i, is estimated from the normal theory–based confidence limits

$$\sigma_{ii} = \left[\left\{\log(u_b) - \log(l_b)\right\}/(2 \times z_{\alpha/2})\right]^2 \tag{5}$$

where u_b and l_b are, respectively, the upper and lower bounds of the reported relative risks, $\exp(y_i)$, and $z_{\alpha/2}$ denotes the $(1 - \alpha/2)$-level standard normal deviate (e.g., use 1.96 for 95% confidence interval).

Following the method proposed by Greenland and Longnecker (1992), one way to estimate the off-diagonal elements σ_{ij} of $\mathbf{\Sigma}$, with $i \neq j$, is to assume that the correlations between the unadjusted log relative risks are approximately equal to those of the adjusted log relative risks. Here, besides the log relative risks, their variances, and exposure levels, we also need to know for each exposure level the number of cases and the number of controls for case–control data (table 4), or the number of cases for incidence-rate data (table 5), or the number of cases and noncases for cumulative incidence data (table 6)—information usually available from the publication.

Table 4. Summary of case–control data

	x_{01}	x_{11}	$\ldots$	x_{i1}	$\ldots$	x_{n1}	Total
			Exposure levels				
Cases	A_0	A_1	$\ldots$	A_i	$\ldots$	A_n	$M_1 = \sum_{i=0}^{n} A_i$
Controls	B_0	B_1	$\ldots$	B_i	$\ldots$	B_n	$M_0 = \sum_{i=0}^{n} B_i$
Total	N_0	N_1	$\ldots$	N_i	$\ldots$	N_n	$M_1 + M_0$

The off-diagonal elements of $\mathbf{\Sigma}$ can be estimated using the following three-step procedure, where formulas used for steps 1 and 2 change according to the study type: case–control, incidence-rate, or cumulative incidence data.

For case–control data, where we model log odds-ratios, the off-diagonal elements σ_{ij} of $\mathbf{\Sigma}$ are computed as follows:

1. Fit cell counts A_i and B_i as modeled in table 4 (which has margin M_1 and N_i), such that

$$(A_i \times B_0)/(A_0 \times B_i) = \exp(y_i) \tag{6}$$

where A_i is the fitted number of cases and B_i is the fitted number of controls at each exposure level (see iterative algorithm described in Greenland and Longnecker 1992, appendix 2).

2. For $i \neq j$, estimate the asymptotic correlation, r_{ij}, of y_i and y_j by

$$r_{ij} = s_0/(s_i s_j)^{1/2} \qquad (7)$$

where $s_0 = (1/A_0 + 1/B_0)$ and $s_i = (1/A_i + 1/B_i + 1/A_0 + 1/B_0)$.

3. Estimate the off-diagonal elements, σ_{ij}, of the asymptotic covariance matrix $\mathbf{\Sigma}$ by

$$\sigma_{ij} = r_{ij} \times (\sigma_i \sigma_j)^{1/2}$$

where σ_i and σ_j are the variances of y_i and y_j, estimated using (5).

The above method can be easily extended to the analysis of incidence-rate and cumulative incidence data, upon redefinition of terms in (6) and (7).

Table 5. Summary of incidence-rate data

	Exposure levels					Total
	x_{01}	x_{11}	...	x_{i1}	...	x_{n1}	
Cases	A_0	A_1	...	A_i	...	A_n	$M_1 = \sum_{i=0}^{n} A_i$
Person-time	N_0	N_1	...	N_i	...	N_n	$M_0 = \sum_{i=0}^{n} N_i$

For instance, for incidence-rate data, where we model log incidence-rate ratios, fit cell counts A_i as modeled in table 5 such that $(A_i \times N_0)/(A_0 \times N_i) = \exp(y_i)$. In (7), we redefine $s_0 = (1/A_0)$ and $s_i = (1/A_i + 1/A_0)$.

Table 6. Summary of cumulative incidence data

	Exposure levels					Total
	x_{01}	x_{11}	...	x_{i1}	...	x_{n1}	
Cases	A_0	A_1	...	A_i	...	A_n	$M_1 = \sum_{i=0}^{n} A_i$
Noncases	B_0	B_1	...	B_i	...	B_n	$M_0 = \sum_{i=0}^{n} B_i$
Total	N_0	N_1	...	N_i	...	N_n	$M_1 + M_0$

Then, for cumulative incidence data, where we model log risk-ratios, fit cell counts A_i as modeled in table 6 such that $(A_i \times N_0)/(A_0 \times N_i) = \exp(y_i)$. In (7), again s_0 and s_1 need to be computed differently: $s_0 = (1/A_0 - 1/N_0)$ and $s_i = (1/A_i - 1/N_i + 1/A_0 - 1/N_0)$.

2.5 Heterogeneity

The analysis of the estimated residual vector $\hat{\mathbf{e}} = \mathbf{y} - \mathbf{Xb}$ is useful to evaluate how close reported and fitted log relative risks are at each exposure level. A statistic for the goodness of fit of the model is

$$Q = (\mathbf{y} - \mathbf{Xb})'\boldsymbol{\Sigma}^{-1}(\mathbf{y} - \mathbf{Xb}) \tag{8}$$

where Q has approximately, under the null hypothesis that the fitted model is correct, a χ^2 distribution with $n - p$ degrees of freedom. If the p-value derived from this statistic is small, we may infer that there is some problem with the model; e.g., perhaps heterogeneity is present or there is some unaccounted-for bias. If, however, the p-value is large, we can conclude only that the test did not detect a problem with the model, not that there is no problem. The Q statistic (like most fit statistics) has low power; i.e., its sensitivity to model problems is limited.

2.6   Log-linear dose–response model for multiple studies

The method discussed in the previous section can be applied to estimate the underlying trend from multiple summarized data. When dealing with multiple studies and multiple exposure levels, a more flexible method of trend estimation requires pooling the study data before estimating the dose–response model (Greenland and Longnecker 1992).

In a meta-analysis of dose–response studies, heterogeneity means that the shape or slope of the dose–response relationship varies among studies (Berlin, Longnecker, and Greenland 1993). The pool-first method increases the number of the log relative risks and dose values available for the analysis and it allows either to get a better fit of the dose–response relationship, by including fractional polynomials and splines in $\mathbf{X}$, or to identify sources of heterogeneity across studies, by including effect modifiers in $\mathbf{X}$.

Fixed-effects dose–response meta-regression model

Let $\mathbf{y}_k$ be the $n_k \times 1$ response vector and let $\mathbf{X}_k$ be the $n_k \times p$ covariates matrix for the kth study, with $k = 1, 2, \ldots, S$. The number of nonreference exposure levels, n_k, for the kth study might vary among the S studies. We pool the data by concatenating the matrices $\mathbf{y}_k$ and $\mathbf{X}_k$

$$\mathbf{y} = \begin{bmatrix} \mathbf{y}_1 \\ \vdots \\ \mathbf{y}_k \\ \vdots \\ \mathbf{y}_S \end{bmatrix} \qquad \mathbf{X} = \begin{bmatrix} \mathbf{X}_1 \\ \vdots \\ \mathbf{X}_k \\ \vdots \\ \mathbf{X}_S \end{bmatrix}$$

so the outcome $\mathbf{y}$ will be an $n \times 1$ vector, where $n = \sum_{k=1}^{S} n_k$, and the linear predictor $\mathbf{X}$ will be an $n \times p$ matrix.

Using the pool-first method, the log-linear model

$$\mathbf{y} = \mathbf{X}\beta + \mathbf{e} \tag{9}$$

becomes a fixed-effects dose–response meta-regression model, where now the vector of random errors, $\mathbf{e}$, has expected value $E(\mathbf{e}) = 0$ and covariance $\text{Cov}(\mathbf{e}) = E(\mathbf{ee}')$ equal to the following symmetric $n \times n$ block-diagonal matrix,

$$\mathbf{\Sigma} = \begin{bmatrix} \mathbf{\Sigma}_1 & & & & \\ \vdots & \ddots & & & \\ \mathbf{0} & & \mathbf{\Sigma}_k & & \\ \vdots & & & \ddots & \\ \mathbf{0} & \ldots & \mathbf{0} & \ldots & \mathbf{\Sigma}_S \end{bmatrix} \tag{10}$$

where $\mathbf{\Sigma}_k$ is the $n_k \times n_k$ estimated covariance matrix for the kth study. We assume that the log relative risks are correlated within each study but uncorrelated across different studies.

The GLS estimators are given by (2) and (3), where the variance–covariance matrix is now given by (10). The summary slope (trend) across studies is a weighted average of each study slope with weighting matrix given by the inverse of $\mathbf{\Sigma}$.

A test for heterogeneity is again given by (8), where the variance–covariance matrix is given by (10). The Q statistic has approximately, under the null hypothesis, a χ^2 distribution with $n - p$ degrees of freedom.

The assumption implicit in a fixed-effects meta-regression model is that each study is estimating the same underlying trend. If heterogeneity is detected then it means that we could fit a better dose–response model, namely, one closer to the observed log relative risks, by either including in the linear predictor transformations of the dose variable and/or interaction terms between exposure dose levels and additional covariates, such as the study design. If important residual heterogeneity is still present after accounting for all known effect modifiers, a random-effects meta-regression dose–response model will be necessary to estimate a summary trend across studies (Berlin, Longnecker, and Greenland 1993).

Random-effects dose–response meta-regression model

We extend the fixed-effects dose–response model (9) to incorporate residual heterogeneity by including an additive random effect

$$\mathbf{y} = \mathbf{X}\beta + \mathbf{Z}\eta + \mathbf{e}$$

where $\mathbf{Z}$ is an $n \times 1$ vector containing the dose variable, first column of $\mathbf{X}$, and η is a random effect with expected value $E(\eta) = 0$ and variance $E(\eta\eta') = \tau^2$, and the random variables η and $\mathbf{e}$ are independent. The τ^2 represents a between-study variance component and quantifies the amount of spread about an overall slope (trend) of the dose variable in the reference category of all covariates specified in $\mathbf{X}$. We estimate the between-study variance using the moment estimator

$$\widehat{\tau}^2 = \frac{Q - (n - p)}{\mathrm{tr}(\mathbf{\Sigma}^{-1}) - \mathrm{tr}\{\mathbf{\Sigma}^{-1}\mathbf{X}(\mathbf{X}'\mathbf{\Sigma}^{-1}\mathbf{X})^{-1}\mathbf{X}'\mathbf{\Sigma}^{-1}\}}$$

where tr denotes the trace of a matrix. A revised variance–covariance matrix, $\mathbf{\Sigma}$, is obtained by replacing the matrices $\mathbf{\Sigma}_k = \mathbf{\Sigma}_k + \widehat{\tau}^2\mathbf{Z}_k\mathbf{Z}_k'$ in the block diagonal matrix (10). The revised matrix $\mathbf{\Sigma}$ is plugged into the GLS estimators $\mathbf{b}$ and $\mathbf{v}$, defined by (2, 3), and into the Q statistic, defined by (8). To get a fully efficient estimator, this procedure is repeated until the difference between successive estimates of $\widehat{\tau}^2$ is less than 10^{-5}. Whenever $\widehat{\tau}^2$ is negative, because $Q < n - p$, it is set to zero. The above iterative GLS method is approximately equivalent to first estimating the slope for each study and then pooling the slopes with a random-effects model (DerSimonian and Laird 1986).

3 The glst command

The estimation command glst is written for Stata 9.1, and it uses several inline Mata functions (see [M-5] **intro**).

3.1 Syntax of glst

glst *depvar dose* [*indepvars*] [*if*] [*in*], <u>se</u>(*stderr*) <u>cov</u>(*n cases*) [[cc | ir | ci]
    <u>pf</u>irst(*id study*) <u>random</u> <u>level</u>(#) eform]

where *depvar*, the outcome variable, contains log relative risks; *dose*, a required covariate, contains the exposure levels; and *indepvars* may contain other covariates, such as transformations of *doses* or interaction terms.

3.2 Options

se(*stderr*) specifies an estimate of the standard error of *depvar*. se() is required.

cov(*n cases*) specifies the variables containing the information required to fit the covariances among correlated log relative risks. At each exposure level, according to the study type, *n* is the number of subjects (controls plus cases) for case–control data (cc); or the total person-time for incidence-rate data (ir); or the total number of persons (cases plus noncases) for cumulative incidence data (ci). The variable *cases* contains the number of cases at each exposure level.

cc specifies case–control data. It is required for trend estimation of a single study unless the option pfirst(*id study*) is specified.

ir specifies incidence-rate data. It is required for trend estimation of a single study unless the option pfirst(*id study*) is specified.

ci specifies cumulative incidence data. It is required for trend estimation of a single study unless the option pfirst(*id study*) is specified.

pfirst(*id study*) specifies the pool-first method with multiple summarized studies. The variable *id* is a numeric indicator variable that takes the same value across correlated log relative risks within a study. The variable *study* must take value 1 for case–control, 2 for incidence-rate, and 3 for cumulative incidence study. Within each group of log relative risks, the first observation is assumed to be the referent.

random specifies the iterative generalized least squares method to estimate a random-effects meta-regression model. Between-study variability of the *dose* coefficient is estimated with the moment estimator.

level(#) specifies the confidence level, as a percentage, for confidence intervals. The default is level(95) or as set by set level; see [U] **20.7 Specifying the width of confidence intervals**.

eform reports coefficient estimates as exp(b) rather than b. Standard errors and confidence intervals are similarly transformed.

3.3 Saved results

glst saves in e():

Scalars

e(N)	number of observations	e(df_gf)	goodness-of-fit degrees of
e(chi2)	model χ^2 statistic		freedom
e(ll)	log likelihood	e(chi2_gf)	goodness-of-fit test
e(tau2)	between-study variance τ^2	e(S)	number of studies
e(df_m)	model degrees of freedom		

Macros

e(cmd)	glst	e(properties)	b V
e(depvar)	name of dependent variable		

Matrices

e(b)	coefficient vector	e(V)	variance–covariance matrix of
e(Sigma)	$\widehat{\Sigma}$ matrix		the estimators

Functions

e(sample)	marks estimation sample

4 Examples

4.1 Case–control data: Alcohol and breast cancer risk

Consider the case–control data shown in table 1 on alcohol and breast cancer (Rohan and McMichael 1988). We use the dataset containing the summarized information, and we calculate the standard errors of the log relative risks from the reported 95% confidence intervals using (5).

```
. use cc_ex

. gen double se = (logub - loglb)/(2*invnormal(.975))
```

We fit the log-linear dose–response model (1) to regress the log relative risks on the exposure level. The command glst fits the covariances and uses the GLS estimator to provide a correct estimate of the linear trend.

```
. glst logrr dose, se(se) cov(n case) cc
Generalized least-squares regression              Number of obs   =        3
Goodness-of-fit chi2(2)    =     1.93             Model chi2(1)   =     4.83
Prob > chi2                =   0.3816             Prob > chi2     =   0.0279
```

logrr	Coef.	Std. Err.	z	P>|z|	[95% Conf. Interval]
dose	.0454288	.0206639	2.20	0.028	.0049284 .0859293

The command glst stores the fitted covariance matrix of the log relative risks in e(Sigma)

```
. matrix list e(Sigma)
symmetric e(Sigma)[3,3]
            c1          c2          c3
r1   .05417235
r2   .01881768   .05627467
r3   .01943145   .02068682   .05632754
```

The exponentiated linear trend for a change of 11 g/d of alcohol level is 1.65 (95% CI = 1.06, 2.57).

```
. lincom dose*11, eform
 ( 1)   11 dose = 0
```

logrr	exp(b)	Std. Err.	z	P>|z|	[95% Conf. Interval]
(1)	1.648255	.3746524	2.20	0.028	1.055709 2.573384

The goodness-of-fit p-value ($Q = 1.93$, $\Pr = 0.3816$) is large. Thus this test detected no problems with the fitted model.

4.2 Incidence-rate data: Fiber intake and coronary heart disease

Consider now the incidence-rate data shown in table 2 on long-term intake of dietary fiber and risk of coronary heart disease among women (Wolk et al. 1999). As we did for case–control data, we use the command glst to get an efficient estimate of the slope.

```
. use ir_ex

. gen double se = (logub - loglb)/(2*invnormal(.975))
```

```
. glst logrr doser, se(se) cov(n case) ir
Generalized least-squares regression              Number of obs  =       4
Goodness-of-fit chi2(3)    =     0.18             Model chi2(1)  =    3.47
Prob > chi2                =  0.9809             Prob > chi2     =  0.0626
```

logrr	Coef.	Std. Err.	z	P>|z|	[95% Conf. Interval]	
doser	-.0232086	.0124649	-1.86	0.063	-.0476394	.0012221

```
. lincom doser*10, eform
 ( 1)   10 doser = 0
```

logrr	exp(b)	Std. Err.	z	P>|z|	[95% Conf. Interval]	
(1)	.7928775	.0988316	-1.86	0.063	.6210185	1.012296

For a 10-g/d increase in total fiber intake, the rate of coronary heart disease decreased by 21% (RR = 0.79, 95% CI = 0.62, 1.01). The linear trend estimated with the `glst` command on summarized data is very close to the linear trend estimated on full data (68,782) reported in the abstract of the paper (RR = 0.81, 95% CI = 0.66, 0.99).

4.3   Cumulative incidence data: High-fat dairy food intake and colorectal cancer risk

Finally, let's consider now the cumulative incidence data shown in table 3 on high-fat dairy food intake and colorectal cancer risk (Larsson, Bergkvist, and Wolk 2005).

```
. use ci_ex
. gen double se = (logub - loglb)/(2*invnormal(.975))
. glst logrr dose, se(se) cov(n case) ci
Generalized least-squares regression              Number of obs  =       4
Goodness-of-fit chi2(3)    =     2.56             Model chi2(1)  =   11.84
Prob > chi2                =  0.4648             Prob > chi2     =  0.0006
```

logrr	Coef.	Std. Err.	z	P>|z|	[95% Conf. Interval]	
dose	-.073636	.0214036	-3.44	0.001	-.1155863	-.0316857

```
. lincom dose*2, eform
 ( 1)   2 dose = 0
```

logrr	exp(b)	Std. Err.	z	P>|z|	[95% Conf. Interval]	
(1)	.8630591	.0369452	-3.44	0.001	.7936024	.9385948

Each increment of two servings per day of high-fat dairy foods corresponded to a 14% reduction in the risk of colorectal cancer (RR = 0.86, 95% CI = 0.79, 0.94). Once again, the linear trend estimated with the `glst` command on summarized data is very close to the linear trend estimated on full data (60,708) reported in the abstract of the paper (RR = 0.87, 95% CI = 0.78, 0.96).

4.4   Meta-analysis: Lactose intake and ovarian cancer risk

Earlier we showed how to estimate a linear trend for a single study. Here we show how to use the command `glst` to estimate a summary linear trend across multiple studies. We consider as a motivating example a meta-analysis of epidemiological studies (six case–control and three cohort studies) investigating the association between lactose intake and ovarian cancer risk (Larsson, Orsini, and Wolk 2006).

Fixed-effects dose–response meta-regression model

We can easily pool trend estimates across studies with the option `pfirst()`, which specifies the variable names identifying the correlated log relative risks and the type of study (case–control or incidence-rate data).

```
. use ma_ex

. glst logrr dose, se(se) cov(n case) pfirst(id study) eform
Fixed-effects dose-response model          Number of studies   =         9
Generalized least-squares regression          Number of obs    =        28
Goodness-of-fit chi2(27)    =    40.25           Model chi2(1)  =      1.11
Prob > chi2                 =   0.0486           Prob > chi2     =    0.2925
```

logrr	exb(b)	Std. Err.	z	P>|z|	[95% Conf. Interval]	
dose	1.025822	.0248455	1.05	0.293	.9782636	1.075693

Overall, there is no evidence of association between milk intake (10 g/d) and risk of ovarian cancer (RR = 1.03, 95% CI = 0.98, 1.08). However, the goodness-of-fit test ($Q = 40.25$, Pr = 0.0486) suggests that we should take into account potential sources of heterogeneity. The estimated association of lactose intake with ovarian cancer risk might depend on the study design. Therefore, we create a product (interaction) term between the type of study (1 for incidence-rate and 0 for case–control data) and the dose variable, and we include it in the model. An alternative would be to stratify the meta-analysis by study design.

```
. gen types = study == 2

. gen doseXtypes = dose*types
```

(*Continued on next page*)

```
. glst logrr dose doseXtypes, se(se) cov(n case) pfirst(id study)
Fixed-effects dose-response model                    Number of studies   =      9
Generalized least-squares regression                    Number of obs    =     28
Goodness-of-fit chi2(26)    =    30.55                   Model chi2(2)    =  10.80
Prob > chi2                 =    0.2453                  Prob > chi2      = 0.0045
```

logrr	Coef.	Std. Err.	z	P>|z|	[95% Conf. Interval]
dose	-.0340478	.0308599	-1.10	0.270	-.094532 .0264365
doseXtypes	.1550466	.0497982	3.11	0.002	.0574439 .2526492

```
. lincom dose + doseXtypes*0, eform
 ( 1)   dose = 0
```

logrr	exp(b)	Std. Err.	z	P>|z|	[95% Conf. Interval]
(1)	.9665253	.0298269	-1.10	0.270	.9097986 1.026789

```
. lincom dose + doseXtypes*1, eform
 ( 1)   dose + doseXtypes = 0
```

logrr	exp(b)	Std. Err.	z	P>|z|	[95% Conf. Interval]
(1)	1.128624	.0441106	3.10	0.002	1.045397 1.218476

No association between milk intake and risk of ovarian cancer was found among six case–control studies ($RR = 0.97$, 95% CI $= 0.91$, 1.03). A positive association between milk intake and risk of ovarian cancer was found among three cohort studies ($RR = 1.13$, 95% CI $= 1.05$, 1.22). A systematic difference in slopes related to study design might result, for instance, from the existence of recall bias in the case–control studies that would not be present in the cohort studies. Now the goodness-of-fit test ($Q = 30.55$, $Pr = 0.2453$) detects no further problems with the fitted model.

Random-effects dose–response meta-regression model

We can also check residual heterogeneity across linear trend estimates by fitting a random-effects model.

```
. glst logrr dose doseXtypes, se(se) cov(n case) pfirst(id study) random
Random-effects dose-response model                   Number of studies   =      9
Iterative Generalized least-squares regression          Number of obs    =     28
Goodness-of-fit chi2(26)    =    28.37                   Model chi2(2)    =   7.29
Prob > chi2                 =    0.3407                  Prob > chi2      = 0.0261
```

logrr	Coef.	Std. Err.	z	P>|z|	[95% Conf. Interval]
dose	-.0443064	.0394422	-1.12	0.261	-.1216116 .0329988
doseXtypes	.1654426	.063171	2.62	0.009	.0416297 .2892555

```
Moment-based estimate of between-study variance of the slope: tau2 =    0.0026
```

The trend estimates for case–control and cohort studies are quite close to the previous ones under fixed-effects models. The between-study standard deviation is close to zero ($\hat{\tau} = 0.0026^{1/2} = 0.05$), which implies that the study-specific trends have only a small spread around the average trend (-0.044) for case–control studies. Furthermore, if we model heterogeneity directly with a random-effects model, without considering any effect modifiers, the results of the meta-analysis briefly described above could not be achieved at all.

```
. glst logrr dose, se(se) cov(n case) pfirst(id study) eform random
Random-effects dose-response model          Number of studies  =       9
Iterative Generalized least-squares regression   Number of obs  =      28
Goodness-of-fit chi2(27)   =    32.17       Model chi2(1)   =    0.20
Prob > chi2                =   0.2259       Prob > chi2     =  0.6519
```

logrr	exb(b)	Std. Err.	z	P>|z|	[95% Conf. Interval]
dose	1.016753	.0374417	0.45	0.652	.9459546 1.092851

```
Moment-based estimate of between-study variance of the slope: tau2  =   0.0059
```

We would simply conclude that, overall, there is no association between lactose intake on ovarian cancer risk (RR $= 1.02$, 95% CI $= 0.95$, 1.09).

5 Empirical comparison of the WLS and GLS estimates

Here we compare and evaluate the uncorrected (WLS) and corrected (GLS) estimates of the linear trend, b, its standard error, se $= \sqrt{v}$, and the heterogeneity statistic, Q. Table 7 summarizes the results for single (sections 4.1–4.3) and multiple studies (section 4.4)

Table 7. Empirical comparison of GLS and WLS estimates

	GLS			WLS			Difference (%)		
	b	se	Q	b	se	Q	b	se	Q
Single study									
Case–control	0.045	0.021	1.93	0.033	0.019	1.72	26.4	9.5	10.5
Incidence-rate	-0.008	0.006	1.61	-0.007	0.004	0.93	14.6	33.7	42.2
Cumulative									
incidence	-0.073	0.021	2.57	-0.098	0.018	2.20	-33.2	15.6	14.1
Multiple									
studies									
Case–control	-0.034	0.031	24.02	-0.042	0.026	30.48	-23.1	17.2	-26.9
Incidence-rate	0.121	0.039	6.54	0.142	0.033	3.24	-17.0	15.0	50.5
Overall	0.025	0.024	40.25	0.026	0.020	52.90	-3.2	16.4	-31.4

The relative differences, expressed as percentages, between the GLS and WLS estimates are calculated as $(\text{GLS} - \text{WLS})/\text{GLS} \times 100$. The GLS estimates of the linear trend, b, could be higher or lower than the WLS estimates, and the small differences are not surprising because both estimators are consistent (Greenland and Longnecker 1992). The Q statistic based on GLS estimates could be higher or lower than the one based on WLS estimates. In the WLS procedure the off-diagonal elements of $\boldsymbol{\Sigma}$, covariances among log relative risks, are set to zeros, whereas in the GLS the covariances are not zeros (see section 2.4). Therefore, the weighting matrix, $\boldsymbol{\Sigma}^{-1}$, in the Q statistic depends both on variances and covariances of the log relative risks. As expected, the GLS estimates of the standard errors, se, are always higher than the WLS estimates of the standard errors for single and multiple studies. The underestimation of the standard error of the uncorrected WLS method somewhat overstates the precision of the trend estimate. Further empirical comparisons between the corrected and uncorrected methods can be found in Greenland and Longnecker (1992).

6   Conclusion

We presented a command, `glst`, to efficiently estimate the trend from summarized epidemiological dose–response data. As shown with several examples, the method can be applied for published case–control, incidence-rate, and cumulative incidence data, from either a single study or multiple studies. In the latter case, the command `glst` fits fixed-effects and random-effects meta-regression models to allow a better fit of the dose–response relation and the identification of sources of heterogeneity. Adjusting the standard error of the slope for the within-study covariance is just one of the statistical issues arising in the synthesis of information from different studies. Other important issues, not considered in this paper, are the exposure scale, publication bias, and methodologic bias (Berlin, Longnecker, and Greenland 1993; Shi and Copas 2004; Greenland 2005). A limitation of the method proposed by Greenland and Longnecker (1992) is the assumption that the correlation matrices of the unadjusted and adjusted log relative risks are approximately equal. In future developments of the command, upper and lower bounds of the covariance matrix will be implemented to assess the sensitivity of the GLS estimators, as pointed out by Berrington and Cox (2003).

7   References

Berlin, J. A., M. P. Longnecker, and S. Greenland. 1993. Meta-analysis of epidemiologic dose–response data. *Epidemiology* 4: 218–228.

Berrington, A., and D. R. Cox. 2003. Generalized least squares for the synthesis of correlated information. *Biostatistics* 4: 423–431.

DerSimonian, R., and N. Laird. 1986. Meta-analysis in clinical trials. *Controlled Clinical Trials* 7: 177–188.

Greenland, S. 1987. Quantitative methods in the review of epidemiologic literature. *Epidemiologic Reviews* 9: 1–30.

———. 2005. Multiple-bias modelling for analysis of observational data (with discussion). *Journal of the Royal Statistical Society, Series A* 168: 267–306.

Greenland, S., and M. P. Longnecker. 1992. Methods for trend estimation from summarized dose–reponse data, with applications to meta-analysis. *American Journal of Epidemiology* 135: 1301–1309.

Grizzle, J. E., C. F. Starmer, and G. G. Koch. 1969. Analysis of categorical data by linear models. *Biometrics* 25: 489–504.

Larsson, S. C., L. Bergkvist, and A. Wolk. 2005. High-fat dairy food and conjugated linoleic acid intakes in relation to colorectal cancer incidence in the Swedish Mammography Cohort. *American Journal of Clinical Nutrition* 82: 894–900.

Larsson, S. C., N. Orsini, and A. Wolk. 2006. Milk, milk products and lactose intake and ovarian cancer risk: A meta-analysis of epidemiological studies. *International Journal of Cancer* 118: 431–441.

Rohan, T. E., and A. J. McMichael. 1988. Alcohol consumption and risk of breast cancer. *International Journal of Cancer* 41: 695–699.

Shi, J. Q., and J. B. Copas. 2004. Meta-analysis for trend estimation. *Statistics in Medicine* 23: 3–19.

Wolk, A., J. E. Manson, M. J. Stampfer, G. A. Colditz, F. B. Hu, F. E. Speizer, C. H. Hennekens, and W. C. Willett. 1999. Long-term intake of dietary fiber and decreased risk of coronary heart disease among women. *Journal of the American Medical Association* 281: 1998–2004.

The Stata Journal (2009)
Forthcoming

Meta-analysis with missing data

Ian R. White
MRC Biostatistics Unit
Cambridge, UK
ian.white@mrc-bsu.cam.ac.uk

Julian P. T. Higgins
MRC Biostatistics Unit
Cambridge, UK
julian.higgins@mrc-bsu.cam.ac.uk

Abstract. A new command, `metamiss`, performs meta-analysis with binary outcomes when some or all studies have missing data. Missing values can be imputed as successes, as failures, according to observed event rates, or by a combination of these according to reported reasons for the data being missing. Alternatively, the user can specify the value of, or a prior distribution for, the informative missingness odds ratio.

Keywords: metamiss, meta-analysis, missing data

1 Introduction

Just as missing outcome data present a threat to the validity of any research study, so they present a threat to the validity of any meta-analysis of research studies. Typically, analyses assume that the data are missing completely at random or missing at random (MAR) (Little and Rubin 2002). If the data are not MAR (i.e., they are informatively missing) but are analyzed as if they were missing completely at random or MAR, then nonresponse bias typically occurs. The threat of bias carries over to meta-analysis, where the problem can be compounded by nonresponse bias applied in a similar way in different studies.

Many methods for dealing with missing outcome data require detailed data for each participant. Dealing with missing outcome data in a meta-analysis raises particular problems because limited information is typically available in published reports. Although a meta-analyst would ideally seek any important but unreported data from the authors of the original studies, this approach is not always successful, and it is uncommon to have access to more than group-level summary data at best. We therefore address the meta-analysis of summary data, focusing on the case of an incomplete binary outcome.

A central concept is the informative missingness odds ratio (IMOR), defined as the odds ratio between the missingness, M, and the true outcome, Y, within groups (White, Higgins, and Wood 2008). A value of 1 indicates MAR, while IMOR $= 0$ means that missing values are all failures, and IMOR $= \infty$ means that missing values are all successes. We allow the IMOR to differ across groups and across subgroups of individuals defined by reasons for missingness, or to be specified with uncertainty.

We will describe `metamiss` in the context of a meta-analysis of randomized controlled trials comparing an "experimental group" with a "control group", but it could be used in any meta-analysis of two-group comparisons. `metamiss` only prepares the data for each study, and then it calls `metan` to perform the meta-analysis. It allows two main types of methods: imputation methods and Bayesian methods.

First, `metamiss` offers imputation methods as described in Higgins, White, and Wood (2008). Missing values can be imputed as failures or as successes; using the same rate as in the control group, the same rate as in the experimental group, or the same rate as in their own group; or using IMORs. When reasons for missingness are known, a mixture of the methods can be used.

Second, `metamiss` offers Bayesian methods that allow for user-specified uncertainty about the missingness mechanism (Rubin 1977; Forster and Smith 1998; White, Higgins, and Wood 2008). These use the prior $\mathrm{logIMOR}_{ij} \sim N(m_{ij}, s_{ij}^2)$ in group $j = E, C$ of study i, with $\mathrm{corr}(\mathrm{logIMOR}_{iE}, \mathrm{logIMOR}_{iC}) = r$.

The approach of Gamble and Hollis (2005) is also implemented. In this approach, two extreme analyses are performed for each study, regarding all missing values as successes in one group and failures in the other. The two 95% confidence intervals are then combined (together with intermediate values), and a modified standard error is taken as one quarter the width of this combined confidence interval. This method appears to overpenalize studies with missing data (White, Higgins, and Wood 2008), but it is included here for comparison.

2 metamiss command

2.1 Syntax

`metamiss` requires six variables (rE, fE, mE, rC, fC, and mC), which specify the number of successes, failures, and missing values in each randomized group. There are four syntaxes described below.

Simple imputation

`metamiss` *rE fE mE rC fC mC*, *imputation_method* [*imor_option*
    imputation_options meta_options]

where

> *imputation_method* is one of the imputation methods listed in section 2.2, specified without an argument.

> *imor_option* is either `imor(` *# | varname* [*# | varname*] `)` or `logimor(` *# | varname* [*# | varname*] `)` (see section 2.3).

> *imputation_options* are any of the options described in section 2.4.

meta_options are any of the meta-analysis options listed in section 2.6, as well as any valid option for `metan`, including `random`, `by()`, and `xlabel()` (see section 2.6).

Imputation using reasons

`metamiss` *rE fE mE rC fC mC,* *imputation_method1* *impuation_method2*
    [*imputation_method3* ...] [*imor_option* *imputation_options* *meta_options*]

where

imputation_method1, imputation_method2, etc., are any imputation method listed in section 2.2 except `icab` and `icaw`, specified with arguments to indicate numbers of missing values to be imputed by each method.

imor_option, imputation_options, and *meta_options* are the same as documented in *Simple Imputation*.

Bayesian analysis using priors

`metamiss` *rE fE mE rC fC mC,* <u>sdlogimor</u>(*#* | *varname* [*#* | *varname*])
    [*imor_option* *bayes_options* *meta_options*]

where

imor_option and *meta_options* are the same as documented in *Simple Imputation*.

bayes_options are any of the options described in section 2.5.

Gamble–Hollis analysis

`metamiss` *rE fE mE rC fC mC,* <u>gamble</u>hollis [*meta_options*]

where

`gamblehollis` specifies to use the Gamble–Hollis analysis.

meta_options are the same as documented in *Simple Imputation*.

2.2   imputation_method

For simple imputation, specify one of the following options without arguments. For imputation using reasons, specify two or more of the following options with arguments. The abbreviations ACA, ICA-0, etc., are explained by Higgins, White, and Wood (2008).

aca$\big[$ (# | *varname* $\big[$ # | *varname* $\big]$) $\big]$ performs an available cases analysis (ACA).

ica0$\big[$ (# | *varname* $\big[$ # | *varname* $\big]$) $\big]$ imputes missing values as zeros (ICA-0).

ica1$\big[$ (# | *varname* $\big[$ # | *varname* $\big]$) $\big]$ imputes missing values as ones (ICA-1).

icab performs a best-case analysis (ICA-b), which imputes missing values as ones in the experimental group and zeros in the control group—equivalent to ica0(0 1) ica1(1 0). If rE and rC count adverse events, not beneficial events, then icab will yield a worst-case analysis.

icaw performs a worst-case analysis (ICA-w), which imputes missing values as zeros in the experimental group and ones in the control group—equivalent to ica0(1 0) ica1(0 1). If rE and rC count beneficial events, not adverse events, then icaw will yield a best-case analysis.

icape$\big[$ (# | *varname* $\big[$ # | *varname* $\big]$) $\big]$ imputes missing values by using the observed probability in the experimental group (ICA-pE).

icapc$\big[$ (# | *varname* $\big[$ # | *varname* $\big]$) $\big]$ imputes missing values by using the observed probability in the control group (ICA-pC).

icap$\big[$ (# | *varname* $\big[$ # | *varname* $\big]$) $\big]$ imputes missing values by using the observed probability within groups (ICA-p).

icaimor$\big[$ (# | *varname* $\big[$ # | *varname* $\big]$) $\big]$ imputes missing values by using the IMORs specified by imor() or logimor() within groups (ICA-IMORs).

The default is icaimor if imor() or logimor() is specified; if no IMOR option is specified, the default is aca.

Specifying arguments

Used with arguments, these options specify the numbers of missing values to be imputed by each method. For example, ica0(mfE mfC) icap(mpE mpC) indicates that mfE individuals in group E and mfC individuals in group C are imputed using ICA-0, while mpE individuals in group E and mpC individuals in group C are imputed using ICA-p. If the second argument is omitted, it is taken to be zero. If, for some group, the total over all reasons does not equal the number of missing observations (e.g., if mfE + mpE does not equal mE), then the missing observations are shared between imputation types in the given ratio. If the total over all reasons is zero for some group, then the missing observations are shared between imputation types in the ratio formed by summing overall numbers of individuals for each reason across all studies. If the total is zero for all studies in one or both groups, then an error is returned. Numerical values can also be given: e.g., ica0(50 50) icap(50 50) indicates that 50% of missing values in each group are imputed using ICA-0 and the rest are imputed using ICA-p.

2.3   imor_option

imor(# | *varname* [# | *varname*]) sets the IMORs or (if the Bayesian method is being
   used) the prior medians of the IMORs. If one value is given, it applies to both
   groups; if two values are given, they apply to the experimental and control groups,
   respectively. Both values default to 1. Only one of imor() or logimor() can be
   specified.

logimor(# | *varname* [# | *varname*]) does the same as imor() but on the log scale.
   Thus imor(1 1) is the same as logimor(0 0). Only one of imor() or logimor()
   can be specified.

2.4   imputation_options

w1 specifies that standard errors be computed, treating the imputed values as if they
   were observed. This is included for didactic purposes and should not be used in real
   analyses. Only one of w1, w2, w3, or w4 can be specified.

w2 specifies that standard errors from the ACA be used. This is useful in separating
   sensitivity to changes in point estimates from sensitivity to changes in standard
   errors. Only one of w1, w2, w3, or w4 can be specified.

w3 specifies that standard errors be computed by scaling the imputed data down to
   the number of available cases in each group and treating these data as if they were
   observed. Only one of w1, w2, w3, or w4 can be specified.

w4, the default, specifies that standard errors be computed algebraically, conditional on
   the IMORs. Conditioning on the IMORs is not strictly correct for schemes including
   ICA-pE or ICA-pC, but the conditional standard errors appear to be more realistic
   than the unconditional standard errors in this setting (Higgins, White, and Wood
   2008). Only one of w1, w2, w3, or w4 can be specified.

listnum lists the reason counts for each study implied by the imputation method option.

listall lists the reason counts for each study after scaling to match the number of
   missing values and imputing missing values for studies with no reasons.

listp lists the imputed probabilities for each study.

2.5   bayes_options

sdlogimor(# | *varname* [# | *varname*]) sets the prior standard deviation for log IMORs
   for the experimental and control groups, respectively. Both values default to 0.

corrlogimor(# | *varname*) sets the prior correlation between log IMORs in the experi-
   mental and control groups. The default is corrlogimor(0).

method(gh | mc | taylor) determines the method used to integrate over the distribution
   of the IMORs. method(gh) uses two-dimensional Gauss–Hermite quadrature and is

the recommended method (and the default). method(mc) performs a full Bayesian analysis by sampling directly from the posterior. This is time consuming, so dots display progress, and you can request more than one of the measures or, rr, and rd. method(taylor) uses a Taylor-series approximation, as in section 4 of Forster and Smith (1998), and is faster than the default but typically inaccurate for sdlogimor() larger than one or two.

nip(#) specifies the number of integration points under method(gh). The default is nip(10).

reps(#) specifies the number of Monte Carlo draws under method(mc). The default is reps(100).

missprior(## [##]) and respprior(##) apply when method(mc) is used, but they are unlikely to be much used. They specify the parameters of the beta priors for $P(M)$ and $P(Y \mid M = 0)$: the parameters for the first group are given by the first two numbers, and the parameters for the second group are given by the next two numbers or are the same as for the first group. The defaults are both beta$(1, 1)$.

nodots suppresses the dots that are displayed to mark the number of Monte Carlo draws completed.

2.6   meta_options

or, rr, and rd specify the measures to be analyzed. Usually, only one measure can be specified; the default is rr. However, when using method(mc), all three measures can be obtained for no extra effort, so any combination is allowed. When more than one measure is specified, the formal meta-analysis is not performed, but measures and their standard errors are saved (see section 2.7).

log has the results reported on the log risk-ratio (RR) or log odds-ratio scale.

id(*varname*) specifies a study identifier for the results table and forest plot.

Most other options allowed with metan are also allowed, including by(), random, and nograph.

2.7   Saved results

metamiss saves results in the same way as metan: _ES, _selogES, etc. The sample size, _SS, excludes the missing values, but an additional variable, _SSmiss, gives the total number of missing values. When method(mc) is run, the log option is assumed for the measures or and rr, and the following variables are saved for each measure (logor, logrr, or rd): the ACA estimate, ESTRAW_*measure*; the ACA variance, VARRAW_*measure*; the corrected estimate, ESTSTAR_*measure*; and the corrected variance, VARSTAR_*measure*. If these variables already exist, then they are overwritten.

3   Examples

3.1   Data

We apply the above methods to a meta-analysis of randomized controlled trials comparing haloperidol to placebo in the treatment of schizophrenia. A Cochrane review of haloperidol forms the basis of our data (Joy, Adams, and Lawrie 2006). Further details of our analysis are given in Higgins, White, and Wood (2008).

The main data consist of the variables `author` (the author); `r1`, `f1`, and `m1` (the counts of successes, failures, and missing observations in the intervention group); and `r2`, `f2`, and `m2` (the corresponding counts in the control group).

3.2   Available cases analysis

The following analysis illustrates `metamiss` output, but the same results could in fact have been obtained by using `metan r1 f1 r2 f2, fixedi`:

```
. use haloperidol
. metamiss r1 f1 m1 r2 f2 m2, aca id(author) fixed nograph
**********************************************************************
******** METAMISS: meta-analysis allowing for missing data ********
********               Available cases analysis           ********
**********************************************************************
Measure: RR.
Zero cells detected: adding 1/2 to 6 studies.

(Calling metan with options: label(namevar=author) fixed eform nograph ...)
              Study  |    ES    [95% Conf. Interval]    % Weight
--------------------+-------------------------------------------
Arvanitis           |   1.417    0.891     2.252         18.86
Beasley             |   1.049    0.732     1.504         31.22
Bechelli            |   6.207    1.520    25.353          2.05
Borison             |   7.000    0.400   122.442          0.49
Chouinard           |   3.492    1.113    10.955          3.10
Durost              |   8.684    1.258    59.946          1.09
Garry               |   1.750    0.585     5.238          3.37
Howard              |   2.039    0.670     6.208          3.27
Marder              |   1.357    0.747     2.466         11.37
Nishikawa_82        |   3.000    0.137    65.903          0.42
Nishikawa_84        |   9.200    0.581   145.759          0.53
Reschke             |   3.793    1.058    13.604          2.48
Selman              |   1.484    0.936     2.352         19.11
Serafetinides       |   8.400    0.496   142.271          0.51
Simpson             |   2.353    0.127    43.529          0.48
Spencer             |  11.000    1.671    72.396          1.14
Vichaiya            |  19.000    1.157   311.957          0.52
--------------------+-------------------------------------------
I-V pooled ES       |   1.567    1.281     1.916        100.00
--------------------+-------------------------------------------

   Heterogeneity chi-squared =   27.29 (d.f. = 16) p = 0.038
   I-squared (variation in ES attributable to heterogeneity) =  41.4%

   Test of ES=1 : z=   4.37 p = 0.000
```

The effect size (ES) refers to the RR in this output. For brevity, future listings include only the four largest studies: Arvanitis, Beasley, Marder, and Selman, with 2%, 41%, 3%, and 42% missing data, respectively. Interest therefore focuses on changes in inferences for the Beasley and Selman studies.

3.3   Imputation methods

We illustrate imputing all missing values as zeros, using the weighting scheme w4, which correctly allows for uncertainty (although in ica0, w1 gives the same answers):

```
. metamiss r1 f1 m1 r2 f2 m2, ica0 w4 id(author) fixed nograph
*****************************************************************
******* METAMISS: meta-analysis allowing for missing data *******
********                Simple imputation            *******
*****************************************************************
Measure: RR.
Method: ICA-0 (impute zeros).
Weighting scheme: w4.
Zero cells detected: adding 1/2 to 6 studies.

(Calling metan with options: label(namevar=author) fixed eform nograph ...)
            Study   |    ES    [95% Conf. Interval]   % Weight
--------------------+-------------------------------------------------
Arvanitis           |  1.362    0.854    2.172        24.38
Beasley             |  1.429    0.901    2.266        25.01
    (output omitted)

Marder              |  1.357    0.745    2.473        14.75
    (output omitted)

Selman              |  2.429    1.189    4.960        10.42
    (output omitted)

--------------------+-------------------------------------------------
I-V pooled ES       |  1.898    1.507    2.390        100.00
--------------------+-------------------------------------------------

  Heterogeneity chi-squared =  21.56 (d.f. = 16) p = 0.158
  I-squared (variation in ES attributable to heterogeneity) =  25.8%
  Test of ES=1 : z=   5.45 p = 0.000
```

The Beasley and Selman trials have more missing data in the control group, so imputing failures increases their estimated RR, and the pooled RR also increases.

3.4   Impute using known IMORs

Now we assume that the IMOR is 0.5 in each group, that is, that the odds of success in missing data are half the odds of success in observed data.

```
. metamiss r1 f1 m1 r2 f2 m2, icaimor imor(1/2 1/2) w4 id(author) fixed nograph
*******************************************************************
******** METAMISS: meta-analysis allowing for missing data ********
********              Simple imputation                    ********
*******************************************************************
Measure: RR.
Method: ICA-IMOR (impute using IMORs 1/2 1/2).
Weighting scheme: w4.
Zero cells detected: adding 1/2 to 6 studies.
(Calling metan with options: label(namevar=author) fixed eform nograph ...)
                Study   |    ES    [95% Conf. Interval]    % Weight
--------------------+-------------------------------------------------
Arvanitis               |  1.399    0.878    2.227          22.12
Beasley                 |  1.120    0.737    1.700          27.47
    (output omitted)
Marder                  |  1.358    0.746    2.473          13.34
    (output omitted)
Selman                  |  1.743    0.973    3.121          14.11
    (output omitted)
--------------------+-------------------------------------------------
I-V pooled ES           |  1.699    1.365    2.115         100.00
--------------------+-------------------------------------------------

    Heterogeneity chi-squared =  24.63 (d.f. = 16) p = 0.077
    I-squared (variation in ES attributable to heterogeneity) =  35.0%

    Test of ES=1 : z=   4.75 p = 0.000
```

The assumption is intermediate between ACA and ICA-0, and so is the result.

3.5   Impute using reasons for missingness

Most studies indicated the distribution of reasons for missing outcomes. We assigned imputation methods as follows:

- For reasons such as "lack of efficacy" or "relapse", we imputed failures (ICA-0).

- For reasons such as "positive response", we imputed successes (ICA-1).

- For reasons such as "adverse event", "withdrawal of consent", or "noncompliance", we considered that the patient had not received the intervention, and we imputed according to the control group rate ICA-pC, implicitly assuming lack of selection bias.

- For reasons such as "loss to follow-up", we assumed MAR and imputed according to the group-specific rate ICA-p.

Counts for these four groups are given by the variables df1, ds1, dc1, and dg1 for the intervention group, and df2, ds2, dc2, and dg2 for the control group.

In some trials, the reasons for missingness were given for a different subset of participants, for example, when clinical outcome and dropout were reported for different time points. In such a case, `metamiss` applies the proportion in each reason-group to the missing population in that trial. In trials that did not report any reasons for missingness, the overall proportion of reasons from all other trials is used.

```
. metamiss r1 f1 m1 r2 f2 m2, ica0(df1 df2) ica1(ds1 ds2) icapc(dc1 dc2)
> icap(dg1 dg2) w4 id(author) fixed nograph
**********************************************************************
******** METAMISS: meta-analysis allowing for missing data ********
********              Imputation using reasons              ********
**********************************************************************
Measure: RR.
Method: ICA-r combining ICA-0 ICA-1 ICA-pC ICA-p.
Weighting scheme: w4.
Zero cells detected: adding 1/2 to 6 studies.

(Calling metan with options: label(namevar=author) fixed eform nograph ...)
          Study    |    ES    [95% Conf. Interval]    % Weight
--------------------+------------------------------------------------
Arvanitis          |  1.381     0.867     2.201         21.37
Beasley            |  1.349     0.892     2.041         27.10
   (output omitted)

Marder             |  1.368     0.751     2.491         12.91
   (output omitted)

Selman             |  1.767     1.037     3.010         16.36
   (output omitted)

--------------------+------------------------------------------------
I-V pooled ES      |  1.785     1.439     2.214        100.00
--------------------+------------------------------------------------

  Heterogeneity chi-squared =  21.86 (d.f. = 16) p = 0.148
  I-squared (variation in ES attributable to heterogeneity) =  26.8%

  Test of ES=1 : z=   5.27 p = 0.000
```

3.6   Impute using uncertain IMORs

Finally, we allow for uncertainty about the IMORs. In the analysis below, we take a $N(0, 4)$ prior for the log IMORs in each group, with the log IMORs in the two groups being a priori uncorrelated.

(*Continued on next page*)

```
. metamiss r1 f1 m1 r2 f2 m2, sdlogimor(2) logimor(0) w4 id(author) fixed
> nograph
*******************************************************************
******** METAMISS: meta-analysis allowing for missing data ********
********          Bayesian analysis using priors          ********
*******************************************************************
Measure: RR.
Zero cells detected: adding 1/2 to 6 studies.
Priors used:  Group 1: N(0,2^2). Group 2: N(0,2^2). Correlation: 0.
Method: Gauss-Hermite quadrature (10 integration points).

(Calling metan with options: label(namevar=author) fixed eform nograph ...)
              Study     |    ES    [95% Conf. Interval]    % Weight
--------------------+---------------------------------------------------
Arvanitis             |  1.416    0.889      2.257          30.37
Beasley               |  1.085    0.506      2.324          11.36
    (output omitted)
Marder                |  1.350    0.737      2.472          18.04
    (output omitted)
Selman                |  1.596    0.671      3.799           8.77
    (output omitted)
--------------------+---------------------------------------------------
I-V pooled ES         |  1.867    1.444      2.413         100.00
--------------------+---------------------------------------------------

    Heterogeneity chi-squared =  20.93 (d.f. = 16) p = 0.181
    I-squared (variation in ES attributable to heterogeneity) =  23.6%

    Test of ES=1 : z=   4.76 p = 0.000
```

Note how the weight assigned to the Beasley and Selman studies is greatly reduced. Because these studies have estimates below the pooled mean, the pooled mean increases.

4 Details

4.1 Zero cell counts

Like metan, metamiss adds one half to all four cells in a 2×2 table for a particular study if any of those cells contains zero. However, this behavior is modified under methods that impute with certainty (ICA-0, ICA-1, ICA-b, and ICA-w): the certain imputation is performed before metamiss decides whether to add one half. As a result, apparently similar options such as ica1 and logimor(99) differ slightly in the haloperidol data, because the logimor(99) analysis adds one half to six studies with $r2 = 0$, whereas the ica1 analysis does this only for three studies with $r2 + m2 = 0$.

4.2 Formula

For the imputation methods, in a given group of a given study, let r, f, and m be the number of observed successes, failures, and missing observations; let $\widehat{\pi} = r/(r+f)$ be the observed success fraction; and let $N = r + f + m$ be the total count. Let k index reason-groups with counts m_k and IMOR θ_k, so that, for example, a group imputed by ICA-0 has $\theta_k = 0$. Then the estimated success fraction is

$$\widehat{\pi}^* = \frac{1}{N}\left(r + \sum_k \frac{m_k \theta_k \widehat{\pi}}{1 - \widehat{\pi} + \theta_k \widehat{\pi}}\right)$$

with the variance obtained by a Taylor-series expansion (Higgins, White, and Wood 2008).

For the Bayesian methods, let δ_j be the log IMOR in group j. Then

$$\widehat{\pi}_j^*(\delta_j) = \frac{1}{N_j}\left(r_j + \frac{m_j e^{\delta_j} \widehat{\pi}_j}{1 - \widehat{\pi}_j + e^{\delta_j} \widehat{\pi}_j}\right)$$

and, for example, the log risk-ratio is obtained by finding the expectation of

$$\log \widehat{\pi}_E^*(\delta_E) - \log \widehat{\pi}_C^*(\delta_C)$$

over the prior $p(\delta_E, \delta_C)$ by numerical integration. The variance is obtained by combining the variance conditional on $p(\delta_E, \delta_C)$ with the variance over $p(\delta_E, \delta_C)$ (White, Higgins, and Wood 2008).

5 Discussion

We believe that ACA is a suitable starting point for a sensitivity analysis that might encompass, for example, imor(1/2 1/2), imor(1/2 2), sdlogimor(2) corrlogimor(1), and sdlogimor(2) corrlogimor(0) (Higgins, White, and Wood 2008; White, Higgins, and Wood 2008). However, a "best" analysis might use reasons for missingness together with subject matter knowledge to assign suitable IMORs. Future work will explore how to integrate the two approaches.

6 References

Forster, J. J., and P. W. F. Smith. 1998. Model-based inference for categorical survey data subject to non-ignorable non-response. *Journal of the Royal Statistical Society, Series B (Statistical Methodology)* 60: 57–70.

Gamble, C., and S. Hollis. 2005. Uncertainty method improved on best–worst case analysis in a binary meta-analysis. *Journal of Clinical Epidemiology* 58: 579–588.

Higgins, J. P. T., I. R. White, and A. M. Wood. 2008. Imputation methods for missing outcome data in meta-analysis of clinical trials. *Clinical Trials* 5: 225–239.

Joy, C. B., C. E. Adams, and S. M. Lawrie. 2006. Haloperidol versus placebo for schizophrenia. *Cochrane Database of Systematic Reviews* 4: CD003082.

Little, R. J. A., and D. B. Rubin. 2002. *Statistical Analysis with Missing Data.* 2nd ed. Hoboken, NJ: Wiley.

Rubin, D. B. 1977. Formalizing subjective notions about the effect of nonrespondents in sample surveys. *Journal of the American Statistical Association* 72: 538–543.

White, I. R., J. P. T. Higgins, and A. M. Wood. 2008. Allowing for uncertainty due to missing data in meta-analysis - Part 1: Two-stage methods. *Statistics in Medicine* 27: 711–727.

The Stata Journal (2009)
Forthcoming

Multivariate random-effects meta-analysis

Ian R. White
MRC Biostatistics Unit
Cambridge, UK
ian.white@mrc-bsu.cam.ac.uk

Abstract. Multivariate meta-analysis combines estimates of several related parameters over several studies. These parameters can, for example, refer to multiple outcomes or comparisons between more than two groups. A new Stata command, `mvmeta`, performs maximum likelihood, restricted maximum likelihood, or method-of-moments estimation of random-effects multivariate meta-analysis models. A utility command, `mvmeta_make`, facilitates the preparation of summary datasets from more detailed data. The commands are illustrated with data from the Fibrinogen Studies Collaboration, a meta-analysis of observational studies; I estimate the shape of the association between a quantitative exposure and disease events by grouping the quantitative exposure into several categories.

Keywords: mvmeta, mvmeta_make, meta-analysis, individual participant data, observational studies

1   Introduction

Standard meta-analysis combines estimates of one parameter over several studies (Normand 1999). Multivariate meta-analysis is an extension that can combine estimates of several related parameters (van Houwelingen, Arends, and Stijnen 2003). In such work, it is important to allow for heterogeneity between studies, usually by fitting a random-effects model (Thompson 1994).

Multivariate meta-analysis has a variety of applications in randomized controlled trials. The simplest is modeling the outcome separately in each arm of a clinical trial (van Houwelingen, Arends, and Stijnen 2003). Other published applications explore treatment effects simultaneously on two clinical outcomes (Berkey, Anderson, and Hoaglin 1996; Berkey et al. 1998; Riley et al. 2007a,b) or on cost and effectiveness (Pinto, Willan, and O'Brien 2005), and explore combining trials comparing more than one treatment (Hasselblad 1998; Lu and Ades 2004). Further applications have been reviewed by Riley et al. (2007b).

There are also possible applications of multivariate meta-analysis in observational studies. These applications include assessing the shape of the association between a quantitative exposure and a disease, which will be illustrated in this article.

One difficulty in random-effects meta-analysis is estimating the between-studies variance. In the univariate case, this is commonly performed by using the method of DerSimonian and Laird (1986). However, maximum likelihood (ML) and restricted maximum likelihood (REML) methods are alternatives (van Houwelingen, Arends, and

Stijnen 2003); in Stata, they are not available in `metan` but can be obtained from `metareg` (Sharp 1998). This article describes a new command, `mvmeta`, that performs REML and ML estimation in the multivariate case by using a Newton–Raphson procedure. `mvmeta` requires a dataset of study-specific point estimates and their variance–covariance matrix. I also describe a utility command, `mvmeta_make`, that facilitates forming this dataset.

2   Multivariate random-effects meta-analysis with mvmeta

2.1   Syntax

mvmeta *b V* [*if*] [*in*] [, reml ml mm <u>fi</u>xed <u>va</u>rs(*varlist*) corr(*expression*)

   <u>start</u>(*matrix*| *matrix_expression*|mm) <u>show</u>start <u>showchol</u>

   <u>keepmat</u>(*bname Vname*) <u>nou</u>ncertainv eform(*name*) bscorr bscov

   missest(*#*) missvar(*#*) *maximize_options*]

where the data are arranged with one line per study, the point estimates are held in variables whose names start with *b* (excluding *b* itself), the variance of *bx* is held in variable *Vxx*, and the covariance of *bx* and *by* is held in variable *Vxy* or *Vyx* (or the corr() option is specified).

If the dataset includes variables whose names start with *b* that do not represent point estimates, then the vars() option must be used.

2.2   Options

reml, the default, specifies that REML be used for estimation. Specify only one of the reml, ml, mm, or fixed options.

ml specifies that ML be used for estimation. ML is likely to underestimate the variance, so REML is usually preferred. Specify only one of the reml, ml, mm, or fixed options.

mm specifies that the multivariate method-of-moments procedure (Jackson, White, and Thompson Forthcoming) be used for estimation. This procedure is a multivariate generalization of the procedure of DerSimonian and Laird (1986) and is faster than the likelihood-based methods. Specify only one of the reml, ml, mm, or fixed options.

fixed specifies that the fixed-effects model be used for estimation. Specify only one of the reml, ml, mm, or fixed options.

vars(*varlist*) specifies which variables are to be used. By default, all variables *b** are used (excluding *b* itself). The order of variables in *varlist* does not affect the model itself but does affect the parameterization.

corr(*expression*) specifies that all within-study correlations take the given value. This means that covariance variable *Vxy* need not exist. (If it does exist, corr() is ignored.)

start(*matrix* | *matrix_expression* | mm) specifies a starting value for the between-studies variance, except start(mm) specifies that the starting value is computed by the mm method. If start() is not specified, the starting value is the weighted between-studies variance of the estimates, not allowing for the within-study variances; this ensures that the starting value is greater than zero (the iterative procedure never moves away from zero). start(0) uses a starting value of 0.001 times the default. The starting value for the between-studies mean is the fixed-effects estimate.

showstart reports the starting values used.

showchol reports the estimated values of the basic parameters underlying the between-studies variance matrix (the Cholesky decomposition).

keepmat(*bname Vname*) saves the vector of study-specific estimates and the vector of the variance–covariance matrix for study *i* as *bnamei* and *Vnamei*, respectively.

nouncertainv invokes alternative (smaller) standard errors that ignore the uncertainty in the estimated variance–covariance matrix and therefore agree with results produced by procedures such as SAS PROC MIXED (without the ddfm=kr option) and metareg. (Note, however, that the confidence intervals do not agree because mvmeta uses a normal approximation, whereas the other procedures approximate the degrees of freedom of a *t* distribution.)

eform(*name*) exponentiates the reported mean parameters, labeling them *name*.

bscorr reports the between-studies variance–covariance matrix as the standard deviations and reports the correlation matrix. This is the default if bscov is not specified.

bscov reports the between-studies variance–covariance matrix without transformation.

missest(#) specifies the value to be used for missing point estimates; the default is missest(0). This is of minor importance because the variance of these missing estimates is specified to be very large.

missvar(#) is used in imputing the variance of missing point estimates. For a specific variable, the variance used is the largest observed variance multiplied by the specified value. The default is missvar(1E4); this value is unlikely to need to be changed.

maximize_options are any options allowed by ml maximize.

3 Details of mvmeta

3.1 Notation

The data for mvmeta comprise the point estimate, y_i, and the within-study variance–covariance matrix, S_i, for each study $i = 1$ to n.

We assume the model

$$
\begin{aligned}
y_i &\sim N(\mu_i, S_i) \\
\mu_i &\sim N(\mu, \Sigma) \\
\Sigma &= \begin{pmatrix} \tau_1^2 & \kappa_{12}\tau_1\tau_2 & . \\ \kappa_{12}\tau_1\tau_2 & \tau_2^2 & . \\ . & . & . \end{pmatrix}
\end{aligned}
$$

where y_i, μ_i, and μ are $p \times 1$ vectors, and S_i and Σ are $p \times p$ matrices. The within-study variance, S_i, is assumed to be known. Our aim is to estimate μ and Σ.

We set $W_i = (\Sigma + S_i)^{-1}$, noting that this depends on the unknown Σ. If Σ were known (or assumed to be the zero matrix, as in fixed-effects meta-analysis), then we would have

$$
\widehat{\mu} = \left(\sum_i W_i \right)^{-1} \left(\sum_i W_i y_i \right)
$$

3.2   Estimating Σ

Methods proposed for estimating Σ in the multivariate setting include extensions of Cochran's method (Berkey et al. 1998), of the DerSimonian and Laird method (Pinto, Willan, and O'Brien 2005) for diagonal W_i, and of likelihood-based methods (van Houwelingen, Arends, and Stijnen 2003). We use the latter because of their generality and optimality properties. Respectively, the likelihood and restricted likelihood are

$$
-2L = \sum_i \left\{ \log|\Sigma + S_i| + (y_i - \mu)'W_i(y_i - \mu) \right\} + np\log 2\pi
$$

$$
-2RL = -2L + \log\left| \sum_i W_i \right| - p\log 2\pi \tag{1}
$$

where W_i is a function of the unknown Σ, as noted above.

We maximize the (restricted) likelihood with a Newton–Raphson algorithm by using Stata's `ml` procedure. To ensure that Σ is nonnegative definite (for example, in the bivariate case, to ensure that the between-studies variances are nonnegative and that the between-studies correlation lies between -1 and 1), the basic model parameters are taken as the elements of a Cholesky decomposition of Σ (Riley et al. 2007b).

3.3   Saved results

As well as the usual `e()` information, `mvmeta` returns the estimated overall mean in `e(Mu)` and the between-studies variance–covariance matrix, the standard deviation vector, and the correlation matrix in `e(Sigma)`, `e(Sigma_SD)`, and `e(Sigma_corr)`, respectively.

3.4 Files required

mvmeta uses the likelihood program mvmeta_l.ado.

4 A utility command to produce data in the correct format: mvmeta_make

4.1 Syntax

mvmeta_make *regression_command* [*if*] [*in*] [*weight*], by(*by_variable*)
    saving(*savefile*) [replace append names(*bname Vname*) keepmat
    usevars(*varlist*) useconstant esave(*namelist*) nodetails pause
    ppfix(none | check | all) augwt(*#*) noauglist ppcmd(*regcmd*[, *options*])
    hard *regression_options*]

    mvmeta_make performs *regression_command* for each level of *by_variable* and stores the results in *savefile* in the format required by mvmeta. *weight* is any weight allowed by *regression_command*.

4.2 Options

by(*by_variable*) is required; it identifies the studies in which the regression command will be performed.

saving(*savefile*) is required; it specifies to save the regression results to *savefile*.

replace specifies to overwrite the existing file called *savefile*.

append specifies to append the current results to the existing file called *savefile*.

names(*bname Vname*) specifies that the estimated coefficients for variable x are to be stored in variable *bnamex* and that the estimated covariance between coefficients *bnamex* and *bnamey* is to be stored in variable *Vnamexy*. The default is names(y S).

keepmat specifies that the results are also to be stored as matrices. The estimate vector and the covariance matrix for study i are stored as matrices *bnamei* and *Vnamei*, respectively, where *bname* and *Vname* are specified with names().

usevars(*varlist*) identifies the variables whose regression coefficients are of interest. The default is all variables in the model, excluding the constant.

useconstant specifies that the constant is also of interest.

esave(*namelist*) adds the specified e() statistics to the saved data. For example, esave(N ll) saves e(N) and e(ll) as variables _e_N and _e_ll.

`nodetails` suppresses the results of running *regression_command* on each study.

`pause` pauses output after the analysis of each study, provided that `pause on` has been set.

`ppfix(none | check | all)` specifies whether perfect prediction should be fixed in no studies, only in studies where it is detected (the default), or in all studies.

`augwt(#)` specifies the total weight of augmented observations to be added in any study in which perfect prediction is detected (see section 7). `augwt(0)` turns off augmentation but is not recommended. The default is `augwt(0.01)`.

`noauglist` suppresses listing of the augmented observations.

`ppcmd(`*regcmd*[, *options*]`)` specifies that perfect prediction should be fixed by using regression command *regcmd* with options *options* instead of by using the default augmentation procedure.

`hard` is useful when convergence cannot be achieved in some studies. It captures the results of initial model fitting in each study and treats any nonzero return code as a symptom of perfect prediction.

regression_options are any options for *regression_command*.

5   Example 1: Telomerase data

Data from 10 studies of the value of telomerase measurements in the diagnosis of primary bladder cancer were reproduced by Riley et al. (2007b). In the table below, taken from that article, `y1` is logit sensitivity, `y2` is logit specificity, and `s1` and `s2` are their respective standard errors, all estimated from 2×2 tables of true status versus test status.

```
. use telomerase
(Riley's telomerase data)
. format y1 s1 y2 s2 %6.3f
. list, noobs clean
    study       y1       s1       y2       s2
        1    1.139    0.406    3.219    1.020
        2    1.447    0.556    1.299    0.651
        3    1.705    0.272    0.661    0.308
        4    0.470    0.403    3.283    0.588
        5    0.856    0.290    4.920    1.004
        6    1.440    0.371    1.386    0.456
        7    0.187    0.306    3.219    1.442
        8    1.504    0.451    2.197    0.745
        9    1.540    0.636    2.269    0.606
       10    1.665    0.412   -1.145    0.434
. generate S11=s1^2
. generate S22=s2^2
```

5.1 Univariate meta-analysis

We first analyze the data by two univariate meta-analyses:

```
. mvmeta y S, vars(y1) bscov
Note: using method reml
Note: using variable y1
Note: 10 observations on 1 variables
```
   (*output omitted*)

```
                                          Number of obs    =          10
                                          Wald chi2(1)     =       38.52
Log likelihood = -8.7276382               Prob > chi2      =      0.0000
```

	Coef.	Std. Err.	z	P>|z|	[95% Conf.	Interval]
Overall_mean						
y1	1.154606	.1860421	6.21	0.000	.7899701	1.519242

```
Estimated between-studies covariance matrix Sigma:
          y1
y1  .18579341
. mvmeta y S, vars(y2) bscov
Note: using method reml
Note: using variable y2
Note: 10 observations on 1 variables
```
   (*output omitted*)

```
                                          Number of obs    =          10
                                          Wald chi2(1)     =       12.93
Log likelihood = -18.728644               Prob > chi2      =      0.0003
```

	Coef.	Std. Err.	z	P>|z|	[95% Conf.	Interval]
Overall_mean						
y2	1.963801	.5460555	3.60	0.000	.8935515	3.03405

```
Estimated between-studies covariance matrix Sigma:
          y2
y2  2.386426
```

These results agree with SAS PROC MIXED as reported by Riley et al. (2007b), except that the standard errors for the overall means are slightly larger (0.5461 for y2, compared with 0.5414 from SAS). This is because SAS does not, by default, allow for uncertainty in the estimated between-studies variance (SAS Institute 1999). mvmeta's nouncertainv option inverts just the elements of the information matrix relating to the overall mean and agrees with SAS PROC MIXED:

(*Continued on next page*)

```
. mvmeta y S, vars(y2) nouncertainv
Note: using method reml
Note: using variable y2
Note: 10 observations on 1 variables
```
   (*output omitted*)

Alternative standard errors, ignoring uncertainty in V:

	Coef.	Std. Err.	z	P>|z|	[95% Conf. Interval]	
Overall_mean						
y2	1.963801	.5413727	3.63	0.000	.9027297	3.024872

5.2 Multivariate analysis

Because sensitivity and specificity are estimated on separate groups of individuals, their within-study covariance is zero. We could generate a new variable, S12=0, but it is easier to use the corr(0) option:

```
. mvmeta y S, corr(0) bscov
Note: using method reml
Note: using variables y1 y2
Note: 10 observations on 2 variables
Note: corr(0) used for all covariances
```
   (*output omitted*)

```
                                    Number of obs    =         10
                                    Wald chi2(2)     =     159.58
    Log likelihood = -24.415968     Prob > chi2      =     0.0000
```

	Coef.	Std. Err.	z	P>|z|	[95% Conf. Interval]	
Overall_mean						
y1	1.166187	.1863275	6.26	0.000	.8009913	1.531382
y2	2.057752	.5607259	3.67	0.000	.9587493	3.156755

```
Estimated between-studies covariance matrix Sigma:
          y1          y2
y1   .20219111
y2  -.7227506   2.5835381
```

Again these results agree with those of Riley et al. (2007b), except that our standard errors are slightly larger because they allow for uncertainty in the between-studies covariance, Σ.

6 Example 2: Fibrinogen Studies Collaboration data

Fibrinogen Studies Collaboration (FSC) is a meta-analysis of individual data on 154,012 adults from 31 prospective studies with information on plasma fibrinogen and major disease outcomes (Fibrinogen Studies Collaboration 2004). As part of the published analysis, the incidence of coronary heart disease was compared across 10 groups defined by baseline levels of fibrinogen (Fibrinogen Studies Collaboration 2005). That analysis used a fixed-effects model; here we allow for heterogeneity between studies by using a random-effects model, but we reduce the analysis to five groups to avoid presenting lengthy output.

In the first stage of analysis, we start with individual-level data including fibrinogen concentration, `fg`, in five levels. Following standard practice in the analysis of these data (Fibrinogen Studies Collaboration 2005), all analyses are stratified by sex and, for two studies that were randomized trials, by trial arm (variable `tr`). We adjust all analyses for age (variable `ages`), although in practice, more confounders would be adjusted for. We use the `esave(N)` option to record the sample size used in each study in variable `_e_N`.

```
. stset duration allchd

(output omitted)

. xi: mvmeta_make stcox ages i.fg, strata(sex tr) nohr
> saving(FSCstage1) replace by(cohort) usevars(i.fg) names(b V) esave(N)
i.fg            _Ifg_1-5           (naturally coded; _Ifg_1 omitted)
Using coefficients: _Ifg_2 _Ifg_3 _Ifg_4 _Ifg_5

-> cohort==1

        failure _d:  allchd
   analysis time _t:  duration
Iteration 0:   log likelihood = -5223.9564
Iteration 1:   log likelihood = -5135.3888
Iteration 2:   log likelihood = -5129.5633
Iteration 3:   log likelihood =  -5129.551
Refining estimates:
Iteration 0:   log likelihood =  -5129.551

Stratified Cox regr. -- Breslow method for ties

No. of subjects =        14436            Number of obs   =      14436
No. of failures =          603
Time at risk    = 127969.6428
                                          LR chi2(5)      =     188.81
Log likelihood  =    -5129.551            Prob > chi2     =     0.0000
```

_t	Coef.	Std. Err.	z	P>|z|	[95% Conf. Interval]	
ages	.0501925	.0072871	6.89	0.000	.03591	.064475
_Ifg_2	.2523666	.1895222	1.33	0.183	-.11909	.6238233
_Ifg_3	.5317069	.1804709	2.95	0.003	.1779905	.8854233
_Ifg_4	.9464425	.1761563	5.37	0.000	.6011824	1.291703
_Ifg_5	1.400935	.1779354	7.87	0.000	1.052188	1.749682

```
                                                     Stratified by sex tr

-> cohort==2

(output omitted)
```

Here are the data stored for the first 15 of the 31 studies; the data also include covariances `V_Ifg_2_Ifg_3`, etc., which are not displayed to save space. The first row of the data below reproduces the results from the `stcox` analysis given above.

```
. use FSCstage1, clear

. format b* V* %5.3f

. list cohort b_Ifg_2 b_Ifg_3 b_Ifg_4 b_Ifg_5 V_Ifg_2_Ifg_2 V_Ifg_3_Ifg_3,
> clean noobs
      cohort    b_Ifg_2    b_Ifg_3    b_Ifg_4    b_Ifg_5   V_Ifg_~2    ~3_Ifg_3
           1      0.252      0.532      0.946      1.401      0.036       0.033
           2     -0.184     -0.032      0.119      0.567      0.348       0.344
           3      0.001     -0.529     -0.339      0.416      0.375       0.323
           4      0.066      0.184      0.407      0.645      0.058       0.053
           5      0.078      0.406      0.544      1.088      0.101       0.083
           6     -0.113      0.456      0.456      0.875      0.065       0.054
           7     -2.149     -0.264     -0.494      0.169      1.336       0.421
           8     -0.039      0.170      0.420      1.053      0.042       0.038
           9      0.443      0.595      0.922      0.797      0.202       0.175
          10      0.356      1.312      0.628      2.133      1.500       1.170
          11      1.297      1.052      1.421      1.752      0.559       0.542
          12      0.323      0.545      0.681      0.540      0.132       0.122
          13     -0.042      0.509      0.560      0.998      0.088       0.072
          14     -2.667     -2.524     -2.010     -1.767      1.337       0.584
          15      5.946      5.420      6.088      7.057    189.088     189.271

  (output omitted)
```

Note the large parameter estimates and very large variances in study 15, which occur because this study has no events in category 1 of `fg`. Details of how such *perfect prediction* is handled are described in section 7.

Now the second stage of analysis:

```
. mvmeta b V
Note: using method reml
Note: using variables b_Ifg_2 b_Ifg_3 b_Ifg_4 b_Ifg_5
Note: 31 observations on 4 variables

  (output omitted)
                                            Wald chi2(4)     =      139.59
Log likelihood = -79.489126                 Prob > chi2      =      0.0000
```

	Coef.	Std. Err.	z	P>|z|	[95% Conf. Interval]	
Overall_mean						
b_Ifg_2	.1615842	.0796996	2.03	0.043	.005376	.3177925
b_Ifg_3	.3926019	.0878114	4.47	0.000	.2204947	.5647091
b_Ifg_4	.5620076	.0905924	6.20	0.000	.3844497	.7395654
b_Ifg_5	.8973289	.0942603	9.52	0.000	.712582	1.082076

```
Estimated between-studies SDs and correlation matrix:
                SD      b_Ifg_2     b_Ifg_3     b_Ifg_4     b_Ifg_5
b_Ifg_2  .22734097            1  .98953788  .97421937  .70621223
b_Ifg_3  .28611302    .98953788          1  .99657543  .80096928
b_Ifg_4  .30834247    .97421937  .99657543          1  .84773246
b_Ifg_5  .32742861    .70621223  .80096928  .84773246          1
```

It is interesting to compare the estimates with those obtained from four univariate meta-analyses, which can be run by `mvmeta b V, vars(b_Ifg_2)`, etc., and are summarized in table 1.

Table 1. Summary of estimates from four univariate meta-analyses

Group	Univariate			Multivariate						
	$\widehat{\mu}_i$	$\text{se}(\widehat{\mu}_i)$	$\widehat{\tau}_i$	$\widehat{\mu}_i$	$\text{se}(\widehat{\mu}_i)$	$\widehat{\tau}_i$	Correlations $\widehat{\kappa}_{ij}$			
2 vs 1	0.200	0.066	0.134	0.162	0.080	0.227	1			
3 vs 1	0.430	0.073	0.196	0.393	0.088	0.286	0.990	1		
4 vs 1	0.568	0.084	0.263	0.562	0.091	0.308	0.974	0.997	1	
5 vs 1	0.840	0.101	0.363	0.897	0.094	0.327	0.706	0.801	0.848	1

The univariate and multivariate methods give broadly similar point estimates, $\widehat{\mu}_i$, but the multivariate method gives rather larger estimates of three between-studies standard deviations, $\widehat{\tau}_i$, and, consequently, larger standard errors for $\widehat{\mu}_i$. A different choice of reference category would yield the same multivariate results but different univariate results. Of course, the multivariate method also has the advantage of estimating the between-studies correlations.

7 Perfect prediction

7.1 The problem

One difficulty that can occur in regression models with a categorical or time-to-event outcome is *perfect prediction* or *separation* (Heinze and Schemper 2002). In logistic regression, for example, perfect prediction occurs if there is a level of a categorical explanatory variable for which the observed values of the outcome are all one (or all zero); in Cox regression, it occurs if there is a category in which no events are observed. Here, as one or more regression parameters go to plus or minus infinity, the log likelihood increases to a limit and the second derivative of the log likelihood tends to zero.

Stata handles this problem in two ways. Stata first attempts to detect perfect prediction. If successful, it drops the relevant observations and term from the model. However, sometimes (in particular, if perfect prediction is in the reference category of a variable with more than two levels) Stata fails to detect perfect prediction. Here Stata reports very large ML estimates, observes that the variance–covariance matrix is singular, and reports a generalized inverse.

In the meta-analysis context, perfect prediction is likely to occur in some studies and not in others. (In the FSC analysis, it occurred in four studies.) Unfortunately, neither of the above solutions is satisfactory. In the first case, the model fit to a study with perfect prediction differs from that fit to other studies and has fewer parameters, so combination across studies is not meaningful. In the second case, some extremely large coefficients have inappropriately moderate standard errors, so they can have an excessive influence on meta-analytic results.

As an example, we use data from FSC study 15, which has no events in the reference category fg==1:

```
. xi: stcox ages i.fg if cohort==15, nohr
  (output omitted)
No. of subjects =            3134              Number of obs    =       3134
No. of failures =              17
Time at risk    = 9465.954814
                                              LR chi2(5)       =      16.43
Log likelihood  =   -127.22742               Prob > chi2      =     0.0057
```

_t	Coef.	Std. Err.	z	P>|z|	[95% Conf. Interval]	
ages	.0357279	.0263705	1.35	0.175	-.0159573	.087413
_Ifg_2	21.36403	.9147602	23.35	0.000	19.57113	23.15692
_Ifg_3	20.84916	.	.	.	.	.
_Ifg_4	21.50048	.8689028	24.74	0.000	19.79746	23.2035
_Ifg_5	22.47926	.7987255	28.14	0.000	20.91379	24.04473

Perfect prediction has not been detected, and the coefficients are appropriately large but with inappropriately small standard errors.

7.2 Solution: Augmentation

mvmeta_make checks for perfect prediction by checking that 1) all parameters are reported and 2) there are no zeros on the diagonal of the variance–covariance matrix of the parameter estimates. If perfect prediction is detected, mvmeta_make augments the data in such a way as to avoid perfect prediction but gives the added observations a tiny weight to minimize their impact on well-estimated parts of the model.

The augmentation is performed at two design points for each covariate x, defined by letting $x = \bar{x} \pm s_x$ (where $\bar{x}$ and s_x are the study-specific mean and standard deviation of x, respectively) and by fixing other covariates at their mean value. The records added at each design point depend on the form of regression model. For logistic regression, we add one event and one nonevent. For other regression models with discrete outcomes, we add one observation with each outcome level. For survival analyses, we add one event at time $t_{min}/2$ and one censoring at time $t_{max} + t_{min}/2$, where t_{min} and t_{max} are the first and last follow-up times in the study. For a stratified Cox model, the augmentation is performed for each stratum.

A total weight of wp is then shared equally between the added observations, where w is specified by the augwt() option (the default is augwt(0.01)), and p is the number of model parameters (treating the baseline hazard in a Cox model as one parameter). The regression model is then rerun including the weighted added observations. For study 15, this yields

```
No. of subjects =        3134.06              Number of obs    =        3134
No. of failures =          17.03
Time at risk    =     9466.077771
                                              LR chi2(5)       =       16.33
Log likelihood  =     -115.75111              Prob > chi2      =      0.0060
```

_t	Coef.	Std. Err.	z	P>|z|	[95% Conf. Interval]
ages	.0353976	.0263231	1.34	0.179	-.0161948	.08699
_Ifg_2	5.946375	13.75093	0.43	0.665	-21.00495	32.89771
_Ifg_3	5.41975	13.75757	0.39	0.694	-21.54459	32.38409
_Ifg_4	6.088434	13.74965	0.44	0.658	-20.86039	33.03726
_Ifg_5	7.057288	13.74605	0.51	0.608	-19.88448	33.99905

Stratified by sex tr

The coefficients for the _Ifg_* terms are reduced but still large, but their large standard errors now mean that they will not unduly influence the meta-analysis. The coefficient and standard error for ages are barely changed. It is useful to compare the variance–covariance matrix of the parameter estimates before augmentation,

```
ages        _Ifg_2      _Ifg_3      _Ifg_4      _Ifg_5
   ages    .00069444
_Ifg_2    .00156723   .83711768
_Ifg_3            0           0           0
_Ifg_4   -.00185585   .49628548           0   .75596628
_Ifg_5   -.00303957   .49370111           0   .50944939   .64022023
```

with that after augmentation:

```
ages        _Ifg_2      _Ifg_3      _Ifg_4      _Ifg_5
   ages    .00069291
_Ifg_2   -.00309014   189.08811
_Ifg_3   -.00465418   188.76205   189.27067
_Ifg_4   -.00650648   188.77085   188.78488   189.05294
_Ifg_5   -.00768805   188.77649   188.79309   188.81504   188.95394
```

Because the covariances in the latter matrix are large, contrasts between groups 2, 3, 4, and 5 will receive appropriately small standard errors. This study will therefore contribute information about contrasts between groups 2, 3, 4, and 5 to the meta-analysis, but it will contribute no information about contrasts between group 1 and other groups.

A related problem occurs if some study has no observations at all in a particular category. The augmentation algorithm is applied here, too, with the modification that the value s_x, used to define the added design points, is taken as the standard deviation across all studies, because the within-study standard deviation is zero.

(Continued on next page)

8 Discussion

8.1 Difficulties and limitations

The main difficulty that might be encountered in fitting multivariate random-effects meta-analysis models is a nonpositive-definite Σ. However, the parameterization used here ensures that Σ is positive semidefinite and achieves a nonpositive-definite Σ if one or more elements of the Cholesky decomposition approach zero. I have encountered non-convergence of the Newton–Raphson algorithm only when the starting value is $\Sigma = 0$, which is avoided by a suitable nonzero choice of starting values, or when inappropriately handled perfect prediction has led to extreme parameter estimates with small standard errors.

The standard error provided for an REML analysis allows for uncertainty in estimating Σ by inverting the second derivative matrix of the restricted likelihood (1). This is not the standard approach (Kenward and Roger 1997), and its properties require further investigation. Confidence intervals based on a t distribution would be a useful enhancement.

At present, the augmentation routine in `mvmeta_make` effectively ignores any category in which perfect prediction occurs but allows information to be drawn from other categories from that study. A larger augmentation would allow information to be drawn from categories with perfect prediction. For example, if the data consist of 2×2 tables, then standard practice would add 0.5 observations to each cell (Sweeting, Sutton, and Lambert 2004). This amounts to assigning to the augmented observations a total weight equal to the number of parameters, and it is tempting to apply this rule more widely (by using `augment(1)`). However, larger augmentation weights have the undesirable property of not being invariant to reparameterization; for example, a different choice of reference category for the `fg` variable in section 6 would lead to somewhat different results. Larger augmentation is probably best implemented by the user.

There are alternate ways to handle perfect prediction, including various forms of penalized likelihood. The methods of Le Cessie and van Houwelingen (1992) and Verweij and van Houwelingen (1994) have been implemented in Stata by the `plogit` and `stpcox` commands, respectively, and both are currently being updated to allow for perfect prediction (G. Ambler, pers. comm.). The method of Firth (1993) is invariant to reparameterization and is being implemented by the author. When suitable routines become available in Stata, they can be called by the `ppcmd()` option in `mvmeta_make`.

8.2 Comparison to other procedures

All the models considered here can also be fit in SAS PROC MIXED, although some programming effort is required to specify the known within-study variances, S_i. The two approaches are very similar, but by default, SAS produces standard errors that ignore the uncertainty in Σ, and produces confidence intervals by using the t distribution on

$n - 1$ degrees of freedom. Further, SAS optionally provides a standard error adjusted to allow for uncertainty in estimating Σ and provides the approximate degrees of freedom of Kenward and Roger (1997), which has good small-sample properties.

Multivariate meta-analysis models cannot be fit by using existing Stata commands, but univariate models can. `metan` differs from `mvmeta` because it uses DerSimonian and Laird (1986) estimation of the random-effects variance. `metareg` offers the choice of DerSimonian and Laird, ML, or REML estimation, so if run without covariates, it can be compared to `mvmeta`. The original `metareg` (Sharp 1998) used the algorithm of Hardy and Thompson (1996) and did not always find the best solution. Version 2 of `metareg`, by Harbord and Higgins (2008), uses Newton–Raphson maximization via `ml`, and produces the same point estimates as `mvmeta` and the same standard errors as `mvmeta` with the `nouncertainv` option. `metareg` produces confidence intervals that allow for nonnormality of the sampling distributions by using the method of Knapp and Hartung (2003); its `z` option produces confidence intervals that agree with `mvmeta`. Of course, `metareg` also has the enormous advantage of handling meta-regression.

8.3   More than two outcomes

Although `mvmeta` handles several outcomes perfectly well, its computing time increases sharply as the number of outcomes increases. `mvmeta` can even computationally handle situations where there are more quantities of interest than studies ($p > n$); however, fitting such large models can be unwise and results can be untrustworthy.

9   Acknowledgments

I thank the FSC for providing access to their data for illustrative analyses: a full list of the FSC collaborators is given in Fibrinogen Studies Collaboration (2005). I also thank Li Su and Dan Jackson for helping me rediscover the Cholesky decomposition parameterization; Stephen Kaptoge and Sebhat Erquo for helpful comments on the programming; James Roger for help in understanding SAS PROC MIXED; and Patrick Royston and Gareth Ambler for discussions about augmentation and penalized likelihoods.

10   References

Berkey, C. S., J. J. Anderson, and D. C. Hoaglin. 1996. Multiple-outcome meta-analysis of clinical trials. *Statistics in Medicine* 15: 537–557.

Berkey, C. S., D. C. Hoaglin, A. Antczak-Bouckoms, F. Mosteller, and G. A. Colditz. 1998. Meta-analysis of multiple outcomes by regression with random effects. *Statistics in Medicine* 17: 2537–2550.

DerSimonian, R., and N. Laird. 1986. Meta-analysis in clinical trials. *Controlled Clinical Trials* 7: 177–188.

Fibrinogen Studies Collaboration. 2004. Collaborative meta-analysis of prospective studies of plasma fibrinogen and cardiovascular disease. *European Journal of Cardiovascular Prevention and Rehabilitation* 11: 9–17.

———. 2005. Plasma fibrinogen level and the risk of major cardiovascular diseases and nonvascular mortality: An individual participant meta-analysis. *Journal of the American Medical Association* 294: 1799–1809.

Firth, D. 1993. Bias reduction of maximum likelihood estimates. *Biometrika* 80: 27–38.

Harbord, R. M., and J. P. T. Higgins. 2008. Meta-regression in Stata. *Stata Journal* 8: 493–519. (Reprinted in this collection on pp. 70–96.)

Hardy, R. J., and S. G. Thompson. 1996. A likelihood approach to meta-analysis with random effects. *Statistics in Medicine* 30: 619–629.

Hasselblad, V. 1998. Meta-analysis of multitreatment studies. *Medical Decision Making* 18: 37–43.

Heinze, G., and M. Schemper. 2002. A solution to the problem of separation in logistic regression. *Statistics in Medicine* 21: 2409–2419.

Jackson, D., I. R. White, and S. G. Thompson. Forthcoming. Extending DerSimonian and Laird's methodology to perform multivariate random effects meta-analyses. *Statistics in Medicine.*

Kenward, M. G., and J. H. Roger. 1997. Small sample inference for fixed effects from restricted maximum likelihood. *Biometrics* 53: 983–997.

Knapp, G., and J. Hartung. 2003. Improved tests for a random-effects meta-regression with a single covariate. *Statistics in Medicine* 22: 2693–2710.

Le Cessie, S., and J. C. van Houwelingen. 1992. Ridge estimators in logistic regression. *Applied Statistics* 41: 191–201.

Lu, G., and A. E. Ades. 2004. Combination of direct and indirect evidence in mixed treatment comparisons. *Statistics in Medicine* 23: 3105–3124.

Normand, S. L. T. 1999. Meta-analysis: Formulating, evaluating, combining and reporting. *Statistics in Medicine* 18: 213–259.

Pinto, E., A. Willan, and B. O'Brien. 2005. Cost-effectiveness analysis for multinational clinical trials. *Statistics in Medicine* 24: 1965–1982.

Riley, R. D., K. R. Abrams, P. C. Lambert, A. J. Sutton, and J. R. Thompson. 2007a. An evaluation of bivariate random-effects meta-analysis for the joint synthesis of two correlated outcomes. *Statistics in Medicine* 26: 78–97.

Riley, R. D., K. R. Abrams, A. J. Sutton, P. C. Lambert, and J. R. Thompson. 2007b. Bivariate random-effects meta-analysis and the estimation of between-study correlation. *BMC Medical Research Methodology* 7: 3.

SAS Institute. 1999. SAS *OnlineDoc Version Eight*. Cary, NC: SAS Institute. http://www.technion.ac.il/docs/sas/.

Sharp, S. 1998. sbe23: Meta-analysis regression. *Stata Technical Bulletin* 42: 16–22. Reprinted in *Stata Technical Bulletin Reprints*, vol. 7, pp. 148–155. College Station, TX: Stata Press. (Reprinted in this collection on pp. 97–106.)

Sweeting, M. J., A. J. Sutton, and P. C. Lambert. 2004. What to add to nothing? Use and avoidance of continuity corrections in meta-analysis of sparse data. *Statistics in Medicine* 23: 1351–1375.

Thompson, S. G. 1994. Why sources of heterogeneity in meta-analysis should be investigated. *British Medical Journal* 309: 1351–1355.

van Houwelingen, H. C., L. R. Arends, and T. Stijnen. 2003. Advanced methods in meta-analysis: multivariate approach and meta-regression. *Statistics in Medicine* 21: 589–624.

Verweij, P. J. M., and J. C. van Houwelingen. 1994. Penalized likelihood in Cox regression. *Statistics in Medicine* 13: 2427–2436.

Appendix: Further Stata meta-analysis commands

Stata users have written meta-analysis commands that have not, so far, been accepted for publication in the *Stata Journal*. Here are brief descriptions of commands known to the editor at the time of publishing this collection. Readers should note that these commands have not undergone the review process required for publication in the *Stata Journal*. This list is likely to be incomplete, and the editor apologizes to authors of any commands that have been overlooked. For the most up-to-date information on these and other meta-analysis commands, readers are encouraged to check the Stata frequently asked question on meta-analysis:

http://www.stata.com/support/faqs/stat/meta.html

- `metannt` is intended to aid interpretation of meta-analyses of binary data by presenting intervention effect sizes in absolute terms, as the number needed to treat (NNT) and the number of events avoided (or added) per 1,000. The user inputs design parameters, and `metannt` uses the `metan` command to calculate the required statistics. This command is available as part of the `metan` package.

  The NNT is the number of individuals required to experience the intervention in order to expect there to be one additional event to be observed. It is defined as the reciprocal of the absolute value of the risk difference (risk of the outcome in the intervention group minus risk in control).

$$\text{NNT} = \frac{1}{|\text{risk difference}|}$$

  Assuming the event is undesirable, this is termed the *number needed to treat to benefit*. If the intervention arm experiences more events, this is commonly referred to as the *number needed to treat to harm*. Because most meta-analyses are based on ratio measures, the risk difference is calculated based on an assumed value of the risk in the control group. The `metannt` command calculates this by deriving an estimate of the intervention effect (e.g., a risk ratio), applying it to a population with a given outcome event risk, and deriving from this a projected event risk if

249

the population were to receive the intervention. The number of avoided or excess events (respectively) per 1,000 population is the difference between the two event risks multiplied by 1,000. Optionally, a confidence interval is also presented, using the confidence limits for the estimated intervention effect applied to the control group event rate.

- `metaninf` investigates the influence of one study on the overall meta-analysis estimate and shows graphically the results when the meta-analysis estimates are computed, omitting one study in each turn. This command makes repeated calls to the `metan` command for its analyses. It was released in 2001 and was last updated in 2004. It requires the user to provide input in the form needed by `metan`. To install the package, type `ssc install metaninf` in Stata. Articles describing `metainf`, a previous version of the command, were published in the *Stata Technical Bulletin* (Tobias 1999, 2000).

- `midas` provides statistical and graphical routines for undertaking meta-analysis of diagnostic test performance in Stata. Primary data synthesis is performed within the bivariate mixed-effects binary regression modeling framework. Model specification, estimation, and prediction are carried out with `xtmelogit` in Stata 10 or the `gllamm` command in Stata 9 by adaptive quadrature. Using the estimated coefficients and variance–covariance matrices, `midas` calculates summary operating sensitivity and specificity (with confidence and prediction contours in summary receiver operating characteristic space), summary likelihood, and odds ratios. Global and relevant test performance metric-specific heterogeneity statistics are provided. `midas` facilitates extensive statistical and graphical data synthesis and exploratory analyses of heterogeneity, covariate effects, publication bias, and influence. Bayes' nomograms and likelihood-ratio matrices can be obtained and used to guide clinical decision making. The minimum required input data are variables containing the elements of the 2×2 contingency tables (true positives, false positives, false negatives, and true negatives) of test results from each study. To install the package, type `ssc install midas` in Stata.

  Further information on the comprehensive suite of facilities provided by `midas` is available at http://www.sitemaker.umich.edu/metadiagnosis/midas_home. In particular, two presentations given at Stata Users Group meetings are available at http://www.sitemaker.umich.edu/metadiagnosis/presentations and via RePEc at http://econpapers.repec.org/paper/bocasug07/4.htm and http://ideas.repec.org/p/boc/wsug07/1.html.

- `meta_lr` graphs positive and negative likelihood ratios in diagnostic tests. It can do stratified meta-analysis of individual estimates. The user must provide the effect estimates (log positive likelihood ratio and log negative likelihood ratio) and their standard errors. Commands `meta` and `metareg` are used for internal calculations. This is a version 8 command released in 2004. To install the package, type `ssc install meta_lr` in Stata.

- `metaparm` performs meta-analyses and calculates confidence intervals and *p*-values for differences or ratios between parameters for different subpopulations, for data stored in the `parmest` format (Newson 2003). To install the package, type `ssc install metaparm` in Stata.

11 References

Newson, R. 2003. Confidence intervals and p-values for delivery to the end user. *Stata Journal* 3: 245–269.

Tobias, A. 1999. sbe26: Assessing the influence of a single study in the meta-analysis estimate. *Stata Technical Bulletin* 47: 15–17. Reprinted in *Stata Technical Bulletin Reprints*, vol. 8, pp. 108–110. College Station, TX: Stata Press.

———. 2000. sbe26.1: Update of metainf. *Stata Technical Bulletin* 56: 15. Reprinted in *Stata Technical Bulletin Reprints*, vol. 10, p. 72. College Station, TX: Stata Press.

Author index

A

Abrams, K. R...125, 128, 129, 139, 143, 148, 194, 231, 234, 236–238

Adams, C. E......224

Adams, J......115

Ades, A. E......231

Afifi, A. A......44

Alexandersson, A......196

ALSPAC Study Team......82

Altman, D. G......29, 30, 33, 44, 45, 55, 64, 70, 73, 76, 109, 110, 112, 115, 119, 132, 140, 168, 181, 182

Ambler, G......245

Anderson, J. J......231

Angelillo, I. F......160

Antczak-Bouckoms, A......231, 234

Antman, E. M......55, 59, 60

Arends, L. R...181, 184, 196, 231, 232, 234

Aronson, N......181

Audet, A. M......160

B

Bartlett, C......110, 114

Becker, B. J......125, 126

Begg, C. B......115, 139, 140, 151, 154, 155, 168, 174

Benjamini, Y......82

Beresford, S. A. A......98

Bergkvist, L......201, 202, 212

Berkey, C. S...30, 93, 98–101, 231, 234

Berlin, J. A......125, 140, 200, 203, 207, 208, 216

Berrington, A......216

Berry, G......140, 141

Best, N......42

Bezemer, P. D......181

Borenstein, M......124, 134

Bossuyt, P. M. M...181, 182, 184, 186, 194–196

Bouter, L. M......181

Bracken, M. B......5

Bradburn, M. J......29, 30, 33, 55, 64, 73, 112, 119, 132, 168, 182

Breslow, N. E......5, 6, 44

Brewer, T. F......30, 100

Buntinx, F......181

Burdick, E......30, 100

C

Carpenter, J......143, 148

Chalmers, I......110

Chalmers, T. C......55, 59, 60, 160

Chan, A.-W......140

Chanock, S......82

Chew, V......196

Chu, H......194

Colditz, G. A......30, 93, 98–101, 201, 211, 231, 234

Cole, S. R......194

Collins, R......5, 126

Cook, D. G......98

Copas, J. B......203, 216

Cornelius, V......181

Cottingham, J......157, 172

Coveney, J......182, 197

Cox, C......194, 196

Cox, D. R......101, 216

Cox, N. J......121

D

D'Agostino, R. B......20

Davey Smith, G. . . . .
     . . . . . . 7, 70, 82, 109, 114, 115,
     117, 118, 125, 131, 133, 138–
     140, 143, 151, 154, 156, 162,
     168, 174
Day, N. E. 44
de Vet, H. C. W. 181
Deeks, J. J. 27, 29,
     30, 33, 34, 44, 45, 55, 64, 70,
     73, 76, 112, 115, 119, 132, 168,
     181, 182, 184, 194, 196, 197
DerSimonian, R. 6, 89, 92, 98, 209,
     231, 232, 245
Detsky, A. S. 7
Devillé, W. L. 181
Dickersin, K. 125
Dooley, G. 27
Douglas, J. B. 196
DuMouchel, W. H. 92, 98
Duval, S. 131, 134, 165
Dwamena, B. 197

E
Easterbrook, P. J. 125
Edgington, E. S. 65, 66
Egger, M. 7, 36,
     64, 70, 109, 110, 112, 114–118,
     125, 126, 131, 133, 134, 138–
     143, 148, 149, 151, 154, 156,
     162, 168, 174, 181, 182, 184,
     194, 196
El ghormli, L. 82
Engels, E. A. 115
Erquo, S. 245

F
Fibrinogen Studies Collaboration . . 238,
     239, 245
Fine, P. E. M. 30, 39
Fineberg, H. V. 30, 100
Firth, D. 244
Fisher, R. A. 65
Flamm, C. R. 181
Fleiss, J. L. 3, 66
Forster, J. J. 219

Fowler, G. 143

G
Galbraith, R. F. 141, 142
Gamble, C. 219
Garcia-Closas, M. 82
Gatsonis, C. A. 181, 182, 184, 185,
     194–196
Gavaghan, D. . . . 36, 115, 116, 138, 140,
     142
Glas, A. S. 181, 182, 184, 194–196
Glass, G. V. 5
Glasziou, P. P. 114, 140, 141
Gluud, C. 181
Gluud, L. L. 181
Gopalan, R. 125
Green, S. 138, 139, 141, 148, 149
Greenland, S. 4, 5, 36, 113, 200,
     202–205, 207, 208, 216
Grizzle, J. E. 204
Gøtzche, P. C. 140

H
Haahr, M. T. 140
Hackshaw, A. K. 134
Haensel, W. 4
Halligan, S. 181
Hamza, T. H. 181, 184, 196
Harbord, R. M. 38, 39, 55, 64,
     73, 97, 125, 132, 134, 138–140,
     142, 143, 148, 151, 162, 181,
     182, 184, 194, 196, 245
Hardy, R. J. 245
Harris, J. E. 92, 98
Harris, R. J. . . . 55, 64, 73, 132, 134, 151,
     162, 181
Hartung, J. 70, 75, 89, 93, 245
Harville, D. A. 93
Hasselblad, V. 231
Hayes, R. J. 110
Hedges, L. V. 6
Heijenbrok-Kal, M. H. . . . 181, 184, 196
Heinze, G. 241
Heisterkamp, S. H. 181
Held, P. H. 126

Hennekens, C. H. 201, 211
Higgins, J. P. T. . .34, 38, 39, 44, 45, 70,
        71, 73–76, 79, 81, 86, 89, 90,
        93, 97, 138, 139, 141, 148, 149,
        218–220, 222, 224, 229, 245
Hirji, K. F. 44
Hoaglin, D. C. 93, 98–101, 231, 234
Hochberg, Y. 82
Holenstein, F. 110, 114
Hollis, S. 219
Hopewell, S. 181
Hopker, S. W. 74
Hricak, H. 182
Hróbjartsson, A. 140
Hu, F. B. 201, 211
Hunink, M. G. M. 181, 184, 196
Hunter, D. 157, 172

I
Ioannidis, J. P. A. 97, 125
Irwig, L. M. 114, 140, 141, 143, 196
ISIS-4 Collaborative Group 126

J
Jackson, D. 232, 245
Jimenez-Silva, J. 55, 59, 60
Jones, D. R. 65, 125, 128, 129, 139,
        143, 148
Jonkman, J. N. 75, 92, 93
Joy, C. B. 224
Jüni, P. 110, 114

K
Kaptoge, S. 245
Kenward, M. G. 244, 245
Ketteridge, S. 17
Kirkwood, B. R. 32, 46
Knapp, G. 70, 75, 89, 93, 245
Koch, G. G. 204
Kupelnick, B. 55, 59, 60

L
L'Abbé, K. A. 7
Laird, N. . . 6, 89, 92, 98, 209, 231, 232,
        245

Lambert, P. C. 44, 194, 231, 234,
        236–238, 244
Lancaster, T. 143
Larsson, S. C. 201, 202, 212, 213
Lau, J. 55, 59, 60, 97, 115, 181
Law, M. R. 134
Lawlor, D. A. 74, 94
Lawrie, S. M. .224
Le Cessie, S. 244
Lewis, J. 5
Light, R. J. 140, 153, 168
Lijmer, J. G. 181
Lipsey, M. W. 76
Littenberg, B. 184
Little, R. J. A. 218
Longnecker, M. P. . . . 200, 202–205, 207,
        216
Lu, G. 231
Lunn, D. J. 42

M
Macaskill, P. . . . 140, 141, 143, 182, 196
Mallett, S. 181, 197
Manly, B. F. J. 79, 80
Manson, J. E. 201, 211
Mant, D. 143
Mantel, N. 4
Matthews, D. R. 125
Mazumdar, M. 115, 139, 151, 154,
        155, 168, 174
McGaw, B. 5
McMichael, A. J. 201, 210
McQuay, H. J. 114
Minder, C. 7, 114,
        115, 117, 118, 131, 133, 138–
        140, 143, 151, 154, 156, 162,
        168, 174
Moher, D. 138, 141, 148, 149
Montori, V. M. 181
Moore, R. A. 114
Moreno, S. G. .140
Morris, C. N. 75, 89, 93, 104
Morris, R. D. 160
Moses, L. E. 184

Mosteller, F. 30, 55, 59, 60, 93, 98–101, 160, 231, 234

N
Newson, R. 82
Nordic Cochrane Centre 186
Norman, S. L. T. 231

O
O'Brien, B. 231, 234
O'Rourke, K. 7
Olkin, I. 6, 115
Orsini, N. 213

P
Paliwal, P. 181
Palmer, T. M. 140
Peters, J. L. . . . 125, 128, 129, 139, 140, 143, 148
Peto, R. 5, 126
Phillips, A. N. 70
Pickles, A. 191, 192, 197
Pillemer, D. B. 140, 153, 168
Pinto, E. 231, 234
Pocock, S. J. 98
Poole, C. 36, 113

R
Rabe-Hesketh, S. 189–192, 197
Reis, I. M. 44
Reitsma, J. B. . . . 181, 182, 184, 194–196
Reynolds, D. J. M. 114
Riley, R. D. 194, 231, 234, 236–238
Robins, J. 4, 5
Roger, J. H. 244, 245
Rohan, T. E. 201, 210
Rosenthal, R. 6, 65
Rothman, N. 82
Rothstein, H. R. 124, 134
Royston, P. 27, 245
Rubenstein, L. Z. 115
Rubin, D. B. 218, 219
Rushton, L. 125, 128, 129, 139, 143, 148
Rutjes, A. W. S. 181, 182, 184, 194–196

Rutter, C. M. 181, 182, 184, 185, 194–196
Rücker, G. 143, 148

S
Salvan, A. 5
Samson, D. J. 181
SAS Institute 237
Scheidler, J. 182
Schemper, M. 241
Schmid, C. H. 97, 115, 181
Schneider, M. 7, 114, 115, 117, 118, 131, 133, 151, 154, 156, 162, 168, 174
Scholten, R. J. P. M. 181, 182, 184, 194–196
Schulz, K. F. 110
Schwarzer, G. 143, 148
Segal, M. R. 182
Shaffer, J. P. 82
Shapiro, D. 184
Sharp, S. J. . . . 35, 64, 67, 70, 73–75, 87, 92–94, 101, 104, 116, 171, 176, 232, 245
Shi, J. Q. 203, 216
Sidik, K. 75, 92, 93
Siegmund, D. 82
Silagy, C. 17, 143
Simes, R. J. 110
Sinclair, J. C. 5
Siu, A. L. 115
Skrondal, A. 189–192, 197
Sleight, P. 5
Smith, M. L. 5
Smith, P. W. F. 219
Snell, E. J. 101
Speizer, F. E. 201, 211
Spiegelhalter, D. J. 42, 86
Stampfer, M. J. 201, 211
Starmer, C. F. 204
Stead, L. 143
Steichen, T. J. . . . 20, 112, 115, 121, 138, 149, 162, 168, 181
Stern, J. M. 110

Sterne, J. A. C. . . . .
. . . . .35, 36, 46, 55, 64, 67, 70,
73, 82, 101, 105, 110, 112, 114–
116, 125, 126, 132, 134, 138–
143, 148, 149, 151, 162, 168,
171, 176, 181, 182, 184, 194,
196
Stijnen, T. 181, 184, 196, 231, 232,
234
Storey, J. D. 82
Stuck, A. E. 115
Su, L. 245
Subak, L. 182
Sutton, A. J. 44, 93, 124, 125, 128,
129, 134, 139, 140, 143, 148,
194, 231, 234, 236–238, 244
Sweeting, M. J. 44, 244

T
Tatsioni, A. 181
Taylor, J. E. 82
Teo, K. K. 126
Terrin, N. 115
Thomas, A. 42
Thompson, J. R. 194, 231, 234,
236–238
Thompson, S. G. 34,
38, 42, 44, 45, 70, 71, 73–76,
79, 81, 86, 89, 90, 92–94, 101,
104, 105, 116, 231, 232, 245
Tramer, M. R. .114
Tweedie, R. 131, 134, 165

V
van der Windt, D. A. W. M. 181
van Houwelingen, H. C. . . . 231, 232, 234
van Houwelingen, J. C. . . . 181, 184, 196,
244
Verweij, P. J. M. 244

W
Wacholder, S. 82
Wald, N. J. 134
Walter, S. D. 143, 184, 196
Warn, D. E. 42

Weintraub, M. 20
Westfall, P. H. 80, 82
White, I. R. 105, 218–220, 222, 224,
229, 232
Whitehead, A. 142
Whitehead, J. 142
Whiting, P. 181, 182, 184, 194, 196
Wieland, G. D. .115
Willan, A. 231, 234
Willett, W. C. 201, 211
Wilson, D. B. 76
Wilson, M. E. 30, 100
Wolk, A. 201, 202, 211–213
Wood, A. M. . . . 218–220, 222, 224, 229
Woolf, B. 141

Y
Young, S. S. 80, 82
Yu, K. K. 182
Yusuf, S. 5, 126

Z
Zarin, D. A. 181
Zwinderman, A. H. 181, 182, 184,
186, 194–196

Command index

C

confunnel command.........124–137

F

funnel command...............15, 20

G

glst command................200–217

L

labbe command.............16–17, 20

M

meta command...............101–102
meta_lr command.................250
metabias command..........138–164
metacum command..............55–64
metafunnel command........109–123,
      126–127, 145
metamiss command..........218–230
metan command........3–54, 61, 120,
      144–145
metandi command...........181–199
    predict after........188–191, 193
metandiplot command.......186–188,
      192–193
metaninf command 250
metannt command...........249–250
metap command................65–68
metaparm command..........250–251
metareg command.............70–106
metatrim command..........165–177
midas command..................250
mvmeta command............231–247
mvmeta_make command...235–236, 239